Tropical Climatology

Tropical Climatology

An Introduction to the Climates of the Low Latitudes

S. Nieuwolt

Professor of Geography, Kenyatta University, Kenya

JOHN WILEY & SONS

London : New York : Sydney : Toronto

Copyright © 1977, by John Wiley & Sons, Ltd.

Library of Congress Cataloging in Publication Data:

Nieuwolt, S.
  Tropical climatology.

  1. Tropics—Climate. 1. Title.
QC993.5.N5   551.6′9′13   76–13454

ISBN 0 471 99406 5

Typeset by Preface (Graphics) Ltd.,
and printed at The Gresham Press, Old Woking, Surrey.

Preface

This book is the result of my experiences during almost fifteen years of teaching climatology to geography students at universities in the tropics. These students usually complete a course in elementary climatology during their first year of studies, and then need a course to provide them with a better understanding of their own climatic environment. Most textbooks of climatology, written by authors of the mid-latitudes and intended mainly for students in the same parts of the world, contribute relatively little material for this type of course, and the relevant literature is often difficult to collect. There are some excellent textbooks on tropical meteorology, but these are too mathematical for many students of geography, especially those with an arts background. Moreover, they lack the geographical viewpoint, which concentrates on the results of the physical processes in the atmosphere, and their importance to mankind, rather than on these processes themselves.

This book aims to fill the gap described above for students of geography and their lecturers all over the world. It assumes that its users have a basic knowledge of general climatology. Because many students have only a limited knowledge of mathematics and physics, the treatment is non-mathematical and formulae have been avoided. For the benefit of students for whom the book is intended, and to aid their lecturers, quantitative data are not given in tables but in the form of diagrams, which are much more effective for use in the classroom.

The book may also be useful to many visitors to the tropics who are going to work there. The increased aid to the tropical countries from the more developed parts of the world reflects the growing realization that the tropics, with nearly half of the world's population, will play a decisive role in the future of our planet and its inhabitants. Economists, planners, agriculturists and engineers from the mid-latitudes, who come to the tropics in connection with these aid programmes, will soon appreciate that many problems of social, technical and economic development are directly or indirectly related to climatic conditions. A good understanding of the tropical climates may therefore help to direct the aid to the areas where it will do most good and to avoid the costly mistakes which have occurred in the past.

I wish to express my gratitude to many of my students who, by their often persistent questions, forced me to delve deeper into the climatic material. They helped me to discard a number of obvious, but essentially wrong, explanations and theories. I am also most grateful to Professor S. Gregory, University of Sheffield, who read the manuscript and corrected a large number of mistakes. Those which remain are, of course, entirely my own responsibility. Thanks are also due to the Director and the Librarian of the East African Meteorological Department in Nairobi, and their staff, who assisted me in many ways during the final stages of the writing of this book.

Nairobi, December 1975

S. Nieuwolt

Contents

	page
1. Introduction: The Tropics	1
Definition of the term 'tropics'	1
Climatic boundaries of the tropics	1
The importance of the tropics	4
2. Radiation Conditions in the Low Latitudes	6
Insolation and its importance	6
Duration and intensity of insolation	6
Terrestrial radiation	11
The radiation balance	13
3. Temperatures in the Tropics	15
Thermal uniformity	15
Diurnal temperature variations	21
The effects of elevation	25
Physiological temperature	28
4. The General Circulation of the Tropical Atmosphere	34
The Hadley cell	35
The central zone of low pressure and convergent air masses	38
The 'anti-trades'	42
The subtropical highs	43
The trade winds	46
5. Regular Variations of the Tropical Circulation	49
Seasonal variations — the monsoons	49
The Asian monsoons	50
Australian monsoons	56

	page
African monsoons	57
South American monsoons	60
Diurnal variations of the general circulation	60
Sea and land breezes	60
Mountain and valley winds	63
Diurnal pressure variations	65
6. Tropical disturbances	67
Thunderstorms	70
Monsoon depressions	74
Linear systems	75
Easterly waves	77
Tropical cyclones	79
7. Water in the Tropical Atmosphere	85
Evapotranspiration.	85
Humidity of the tropical atmosphere	88
Relative humidity	92
Condensation	94
Tropical clouds	95
Cloudiness	98
8. Tropical Precipitation	102
The origin of tropical rainfall	103
Total rainfall	104
The effects of elevation	107
The variability of annual rainfall over time	107
The seasonal distribution of rainfall	109
The diurnal variation of rainfall	116
Rainfall frequency	119
Rainfall intensity	121
Rainstorms	125
9. Tropical Climates	129
9.1 Tropical monsoon Asia	130
9.1.1 Equatorial monsoon climates	131

9.1.2 Dry and wet monsoon climates   134
9.1.3 The dry parts of tropical monsoon Asia   144
9.2 Tropical Africa   147
9.2.1 Equatorial Africa   150
9.2.2 West Africa and the southern Sudan   152
9.2.3 Southern tropical Africa   154
9.2.4 East Africa   157
9.2.5 Madagascar   160
9.3 Tropical America   162
9.3.1 The Caribbean region   164
9.3.2 Central America   166
9.3.3 Tropical South America   169
9.4 The tropical oceans   176

10. Applied Tropical Climatology   181

Tropical climates and energy production   181
Tropical climates and agriculture   182
Solar radiation   183
Temperature and humidity conditions   186
Rainfall   187
Tropical soils   192
The importance of climatic disadvantages   194
Man's adaptations to climate   194

Author Index   198

Geographical Index   201

Subject Index   205

CHAPTER 1

Introduction: The Tropics

Definition of the Term 'Tropics'

This book deals with the climatic conditions in those parts of the world which are commonly referred to as the 'tropics'. As this term has no exact definition, these pages will start with an indication of which regions are covered in this book. To avoid confusion, the term 'subtropics', which is equally poorly determined, will be avoided, except in some generally used standard expressions.

The word 'tropics' is, of course, derived from the Tropics (with a capital T) of Cancer and Capricorn, the parallels at a latitude of 23½ degrees, which indicate the outer limits of the areas where the sun can ever be in zenith (Figure 1.1). It is generally understood that the tropical areas are mainly found between these two lines. They are therefore the regions of the 'low' latitudes, but the outer limits of these 'low' latitudes are not easily determined, because the Tropics of Cancer and of Capricorn themselves are unsuitable as boundaries. They are too rigid; some regions with clear tropical characteristics are found at latitudes of more than 23½ degrees while, on the other hand, some clearly non-tropical areas are found much closer to the equator.

The best way to determine the outer limits of the tropical areas is therefore to use common characteristics which distinguish these regions from the rest of the world. As latitude is a major factor controlling climatic conditions, the most important of these common features are those of climate. Other typical characteristics of the low latitudes, such as those of vegetation, soil, agriculture and economic development are all, directly or indirectly, related to their common climatic features.

Climatic Boundaries of the Tropics

Undoubtedly, the most important common climatic feature of the low latitudes is the absence of a cold season, as illustrated by the old phrase 'where winter never comes'. To define the tropics by this lack of a winter season cannot be done with a simple temperature limit, as for instance the mean temperature of 18 °C for the coldest month of the year (Köppen, 1936). This method would exclude the tropical highlands, where temperatures frequently remain well below this limit; yet these areas are truly tropical because they experience no winter. They can easily be included in the tropics by using not the actual temperatures but temperatures reduced to sea level (Figure 1.1).

It must be admitted that temperatures reduced to sea level are rather fictitious figures over many continental areas. They also introduce some errors, in that they are based on a standard rate of temperature decrease with elevation, whereas the actual decline varies considerably, both with season and with location (Lautensach and Bögel, 1956). However, on a world scale these errors are relatively small and the boundary itself has many practical advantages: reliable data for its computation are readily available for many stations, interpolation between stations is therefore easy, and its general form is simple.

Another indicator of the absence of a cold season is, of course, the annual range of temperature. This figure shows a general increase with latitude, but it also exhibits a strong influence of continentality. It cannot therefore be used as a realistic boundary of the tropics. But when the annual range is compared to the mean daily range of temperature the situation is different. In the mid-latitudes the annual range generally exceeds the mean daily range, but in the tropics the reverse relation prevails, as indicated by the well-known saying, originating from Alexander von Humboldt: 'the nights are the winter of the tropics'. This is an important climatic characteristic and its outer limit is the line where the annual and daily ranges are about equal (Figure 1.1).

Over the oceans, where the location of a climatic boundary is only approximate anyway, these two limits of the tropics show a general similarity in position, which provides a good indication of the outer limits of the tropics in these areas. But over the continents, where the two lines run farther apart, we shall use the 18 °C sea level isotherm as the outer limit of the tropics. This is done merely as a matter of expediency, this line being simpler in form and easier to interpolate.

Delimited in this way, the tropics constitute a belt around the equator, varying in width from about 40 to 60 degrees in latitude (Figure 1.1).

Having determined the outer limits of the tropics by a climatic boundary, it should be remembered that they are largely arbitrary. Actual climatic conditions differ gradually over long distances and moreover vary a great deal from year to year. The indicated boundaries should therefore be regarded as representing transition zones. The actual limit of the tropics fluctuates from year to year around the average position indicated on the map.

The term 'tropics' is often assumed to include only regions where sufficient rain is received to carry out most forms of crop agriculture without irrigation (Gourou, 1953). Sometimes this condition is expressly stated by the use of the term 'humid tropics', but more often the assumption is made tacitly.

The exact amount of rainfall necessary for crop agriculture cannot be determined easily, because it depends not only on a number of climatic factors, such as temperature, wind speed, sunshine hours and the seasonal distribution of rainfall, but also on conditions of soil, drainage and agricultural methods. A widely accepted estimate based on annual mean temperature and seasonal rainfall regime indicates values between about 450 mm and 600 mm per year

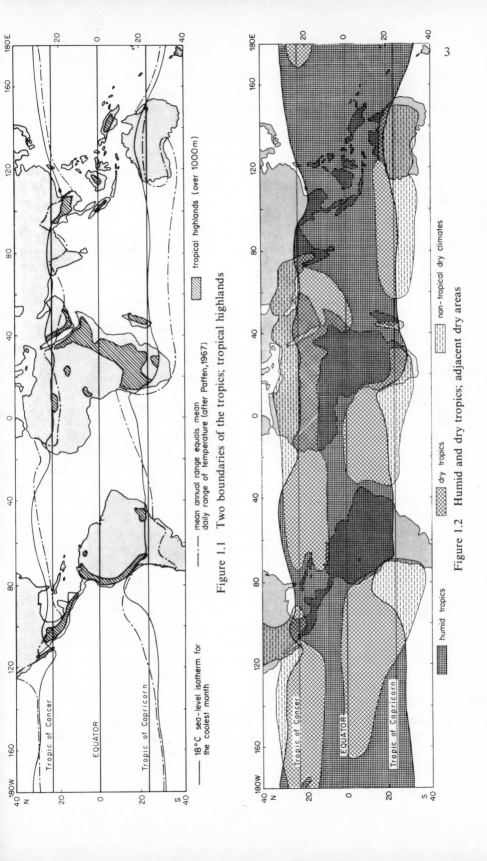

Figure 1.1 Two boundaries of the tropics; tropical highlands

—·— 18°C sea-level isotherm for the coolest month

—— mean annual range equals mean daily range of temperature (after Paffen, 1967)

▨ tropical highlands (over 1000m)

Figure 1.2 Humid and dry tropics; adjacent dry areas

▨ humid tropics   ▨ dry tropics   ▨ non-tropical dry climates

3

(Köppen, 1936). This value is used here to divide the tropics into humid and dry parts (Figure 1.2). The dry tropics, frequently excluded when tropical areas are considered, are included in our treatment in this book because some of them occupy locations rather close to the equator.

However, in these dry areas, and especially near the outer limits of the tropics where rainfall is very low, our temperature boundary of the tropical areas is largely irrelevant. Here, all that matters to man, animal and vegetation is the rainfall, or other sources of water. Temperatures are, in comparison, of little consequence. Therefore the dry tropics belong, in many ways, to the dry climates outside the tropics (Figure 1.2).

There are many other methods of delimiting the tropics, either by climatic elements or by the effects of climate, on, for instance, the natural vegetation. However, our boundaries have the advantage that they can be determined from easily available temperature and rainfall data.

The Importance of the Tropics

The tropics, as delimited in Figure 1.2, occupy about 40 per cent of the land surface of the earth. The total population of these areas amounts to around 1400 million people, which corresponds to 40 per cent of the world population. Almost all these people live in the humid tropics, and more than half of them are concentrated in the tropical areas of southern and eastern Asia.

In economic terms, most tropical countries belong to the group of developing nations, which are characterized by low standards of living, a strong concentration on agriculture and predominance of production of raw materials rather than industrial products. The poverty of the tropical countries is illustrated by the fact that more than two-thirds of them have a Gross National Product below the world median of $310 per capita. And, of the poorest 50 countries of the world, 38 are located in the tropics (*World Bank Atlas*, 1972).

Tropical agriculture is, of course, largely controlled by climatic conditions. It is mainly devoted to the growing of food crops such as rice, maize, cassava, coconuts, bananas, groundnuts and sorghum. Most of these crops are produced on a subsistence basis. Where cash crops are produced, they are often exported to the mid-latitudes, because there are few non-tropical substitutes for such tropical products as coffee, tea, cocoa, palm oil, pepper and many other spices, pineapples and pyrethrum. Other agricultural products of the tropics have a price advantage on the world markets, either because their yields are highest in the tropical climates, or because labour costs are lowest in the tropics. To this category belong natural rubber, cane sugar, cotton, tobacco and groundnuts.

Politically, the tropical countries are much more important than they are economically. This is not solely the result of their numbers (over 50 countries represented in the United Nations are situated in the tropics), their political significance is also the consequence of their common policy; their united stand is frequently independent of the big power blocks. This attitude is largely due to their common history of colonial domination. One of the main origins of colonialism was the desire of the European countries to control the production and trade of tropical agricultural products. Colonial rule was therefore heavily concentrated in the tropics.

Since they have become independent, the tropical countries have usually taken a common and uncompromising attitude against all forms of neo-colonialism in the economic sphere, against racial domination and against the last remnants of colonial rule. This common policy during the last 15 years or so has given this group of nations an important voice in the international political arena.

References

Gourou, P., 1953, *Les pays tropicaux*, Paris, pp. 1, 2.

Köppen, W., 1936, *Das geographische System der Klimate*, Berlin, Gebr., Bornträger, 44 pp.

Lautensach, H. and Bögel, R., 1956, Der Jahresgang des mittleren geographischen Höhengradienten der Lufttemperatur in den verschiedenen Klimagebieten der Erde, *Erdkunde*, **10**, 270–282.

Paffen, K., 1967, Das Verhältnis der Tages- zur jahres-zeitlichen Temperatur-schwankung, *Erdkunde*, **21**, 94–111.

World Bank Atlas 1972, Washington D.C., p. 4.

CHAPTER 2

Radiation Conditions in the Low Latitudes

Insolation and its Importance

Over 99 per cent of all energy in the earth's atmosphere has its origin in radiation from the sun, or *insolation*. A minute fraction of atmospheric energy is supplied by heat from the earth itself, either by volcanic eruptions or by the decay of radioactive minerals and by the burning of organic material: climatologically it has no importance.

All movements and changes in the atmosphere are ultimately caused by variations in the amount of insolation received. These variations can be in time or in place. They are the main cause of climatic differentiation. The main characteristics of the tropical climates, which were described in Chapter 1, have their origin largely in the radiation conditions of the low latitudes.

The amount of insolation received at the outer limit of the atmosphere, at normal incidence and when the sun is at its mean distance from the earth, is about 2 gram calories per square centimetre per minute (Sellers, 1965, p. 11). This is called the *solar constant*. Variations of the solar constant, caused by changes in the radiative activity of the sun, rarely exceed 2 per cent of the average value. In the following parts of this chapter they are therefore disregarded.

The total amount of radiation received at any place on earth depends on two factors: the duration and the intensity of insolation. Both factors are controlled by the movements of the earth: its rotation around its axis and its annual orbit around the sun.

Duration and Intensity of Insolation

The duration of insolation is indicated by length of day, which depends entirely on the earth's rotation around its axis. Because of the angle of 67½ degrees between the axis of rotation and the plane of the earth's orbit around the sun, places on the summer hemisphere enjoy longer days than those having winter; but the total annual exposure to the sun is the same for all places on earth. The difference between summer and winter days increases with distance from the equator, reaching its extreme at the poles, where six months of continuous daylight are followed by six months without it. Only on the equinoxes, March 23rd and September 22nd, are days and nights of equal length everywhere.

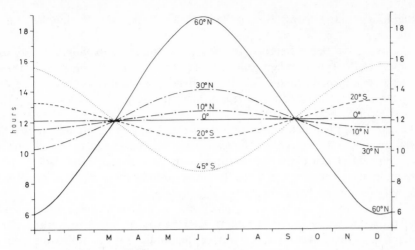

Figure 2.1 Length of day during the course of the year at various latitudes

Conditions at three tropical latitudes (0°, 10° N and 20° S) are compared with three extratropical latitudes (30° N, 45° S and 60° N) to show this effect (Figure 2.1). On this, and on the three following diagrams, latitudes have been chosen from both the northern and southern hemisphere, to emphasize the occurrence of summer and winter conditions in the higher latitudes during opposite periods of the year.

At the equator all days of the year are of equal length, namely twelve hours and seven minutes. Astronomically the duration would be twelve hours exactly, but it takes 3½ minutes for the upper half of the sun to disappear under the horizon at sunset and, similarly, 3½ minutes before the centre of the sun's disc is at the horizon while the upper half of it already provides insolation, at sunrise.

With increasing latitude the difference between the shortest and the longest day of the year grows. In the low latitudes the increase is about 7 minutes per degree, but in the higher latitudes the difference is larger: between 50° and 60° it amounts to about 28 minutes per degree of latitude.

The most important aspect of this factor is clearly illustrated by the diagram: in the low latitudes the seasonal variations in the length of day are insignificant. It is this feature which often surprises newcomers to the tropics: the generally high temperatures are, to people from the mid-latitudes, associated with long summer days. Yet the days in the tropics are always short, the sun rarely sets later than 7.30 p.m. local time.

The *intensity of insolation* is controlled by the movement of the earth around the sun. As the earth's orbit is not a perfect circle, the distance between the earth and the sun varies during the course of the year. On January 3rd, when the earth is nearest to the sun, the intensity of insolation is about 7 per cent higher than on July 4th, when the distance between earth and sun is at its maximum. This difference is the same at all latitudes. It explains a slight asymmetry in diagrams which illustrate the distribution of the intensity of insolation during the course of the year.

Theoretically, the difference in distance from the sun would make the summers of the southern hemisphere hotter, and its winters colder, than those of the northern hemisphere. But these effects are completely masked by the stronger continentality of the northern hemisphere, which causes exactly the opposite conditions.

Climatologically much more important are differences in intensity caused by variations in the sun's elevation. *Elevation* in this respect refers to the position of the sun in the sky; it expresses this position in degrees above the horizon. It is usually indicated for the sun's position at noon local time, which is, of course, the daily maximum elevation.

There are three reasons why a high elevation of the sun causes more intense insolation. The first is that rays coming from the sun in a high position above the horizon are spread over a smaller surface than oblique rays, originating from a low sun. The intensity of insolation is proportional to the sine of the angle of incidence and, for a horizontal part of the earth's surface, this angle is equal to the elevation of the sun.

The second reason is that a high position of the sun means a relatively short passage of the solar radiation through the atmosphere, with consequently lower losses by scattering through atmospheric dust particles. This effect is clearly demonstrated by the fact that to look at the sun at sunrise or sunset with the unprotected human eye is quite harmless, whereas it would be dangerous to do so when the sun is higher in the sky. Scattering affects especially the shorter wavelengths of the sun's radiation, so the longer, reddish rays prevail when scattering is strong. Hence the red colour of a low sun. A similar colour effect can be induced by a large number of dust particles in the atmosphere, as caused by industrial pollution, desert dust storms, volcanic explosions or large forest fires.

A third effect which is influenced by the sun's elevation is the reflection of insolation at the earth's surface. While mainly controlled by the physical characteristics of the surface, especially its colour, under the same circumstances this reflection is strongest when the angle of incidence of the sun is low. Reflection is indicated as the *albedo*, which is the percentage of the incoming radiation that is returned unchanged. The albedo increases when the elevation of the sun diminishes. This effect is of great climatological importance because it is strong over water surfaces, which cover over 70 per cent of the earth's surface (Table 2.1).

The elevation of the sun at noon during the course of the year, at various latitudes, is illustrated in Figure 2.2. The vertical axis of the diagram is divided according to the sine of the elevation. The intensity of insolation is approximately proportional to this value and the diagram therefore indicates relative intensities of insolation. Losses in the atmosphere and at the earth's surface are disregarded.

Table 2.1 Albedo of water surfaces
for different elevations of the sun

Elevation (in degrees)	Albedo (%)
60 and over	2·1
50	2·5
40	3·4
30	6·0
20	13·4
10	34·8
5	58·4
0	100·0

(Source: List, 1958, p. 444.)

The low latitudes are characterized by small seasonal variations and a continuously high intensity. The higher latitudes show large seasonal differences, caused mainly by low intensities during winter. From Figure 2.2 it can be deduced that latitudinal differences in the intensity of insolation are small during summer, but large during the winter.

The combined effects of duration and intensity of insolation show that the long summer days of the higher latitudes more than compensate for the

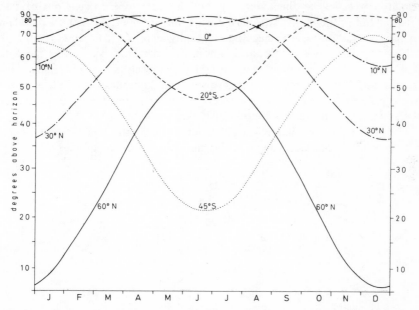

Figure 2.2   Elevation of the sun at noon local time, at the same latitudes as in Figure 2.1 during the course of the year.

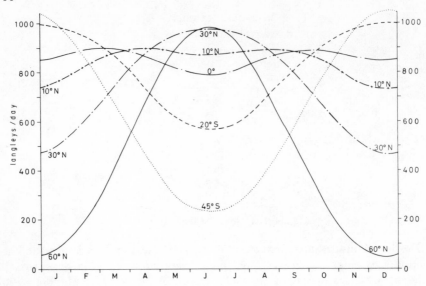

Figure 2.3   Radiation received at the outer limits of the atmosphere (source: List, 1958, p. 418)

relatively low elevations of the sun (Figure 2.3). The highest daily amounts of insolation received at the outer limits of the atmosphere occur not near the equator, but in the higher latitudes, during mid-summer. At that time the total amount of radiation received each day actually increases from the equator to the pole, where the sun is continuously above the horizon. On the other hand, the short days of the winter combined with the very low elevations of the sun reduce the insolation in the higher latitudes to very small amounts during that season. At the poles no insolation at all is received during almost the whole six months of the winter, as the sun is continuously below the horizon.

The tropics receive relatively large amounts of insolation throughout the year, but never reach the high values of the higher latitudes in summer.

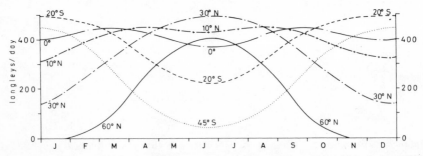

Figure 2.4   Solar radiation received at the earth's surface, assuming an atmospheric transmission coefficient of 0.60 (source: List, 1958, p. 421)

When losses in the atmosphere are considered, the latitudinal differences are smaller (Figure 2.4). For the calculation of this diagram the losses were standardized according to length of path in the atmosphere, assuming a constant transparency at all latitudes (List, 1958, p. 421). The summer maxima in the high latitudes are much more reduced by these losses than the tropical values, derived from higher elevations of the sun. Highest values at the earth's surface are at a latitude of about 30 degrees, during summer, but the difference from the amounts of insolation received near the equator is relatively small.

Actual losses in the atmosphere are largely controlled by cloudiness, but this factor cannot easily be standardized. Generally, cloudiness is high near the equator, low in the dry latitudes of about 20–30 degrees, higher again near the polar front in latitudes between 40 and 60 degrees, and low near the poles. However, differences between various longitudes are considerable, depending on continentality, relief and temperature of the ocean water surfaces.

Similarly, losses by reflection from the earth's surface vary. They are highest over snow surfaces, where up to 80 per cent of the incoming radiation might be reflected. This reduces the net amount of insolation available for heating the earth's surface and its atmosphere, especially at high elevations and in polar areas.

Of the solar radiation received at the outer limits of the atmosphere only about 15 per cent is absorbed directly by the atmosphere. Direct absorption is therefore only a minor source of atmospheric energy. At the earth's surface, however, a much larger proportion of the insolation is absorbed; approximately 47 per cent is turned into heat there. Both these percentage figures are estimates for the whole earth and annual totals. Actual values vary considerably according to the nature of the earth's surface, water surfaces absorbing much more readily than land surfaces. Atmospheric conditions and seasonal differences also cause variations in the absorption of solar radiation, but the general picture is that most of the solar radiation passes through the atmosphere to heat the surface of the earth.

This heating takes place during the day only and results in an increase of temperature. But, in the long run, the earth's surface is not getting hotter, because the energy received from the sun is entirely re-radiated. This is the *terrestrial long-wave or infrared radiation*, characterized by longer wavelengths than the solar radiation.

Terrestrial Radiation

Terrestrial radiation often reduces the temperatures near the earth's surface at night and is therefore sometimes called nocturnal radiation. This term is better avoided, as it implies that this radiation is absent during the day. Actually it always takes place when there is a temperature gradient between the earth's surface and the lower layers of the atmosphere. During the day the terrestrial radiation is usually overshadowed by insolation, but nevertheless it can be present, especially on cloudy days.

Because of its longer wavelengths the terrestrial radiation is readily absorbed by the atmosphere, chiefly by its water vapour and cloud droplets. Only about 3 per cent of the terrestrial radiation is directly lost to outer space.

The heating of the atmosphere is therefore an indirect process: the insolation heats the earth's surface, which in turn heats the atmosphere. Because water vapour and cloud droplets are most prevalent in the lower parts of the atmosphere, the heating process takes place mainly near the earth's surface. The layers which are heated in this way are generally characterized by a decrease of temperature with elevation. Its rate is variable, but normally around 0.5 °C per 100 metres. The lower part of the atmosphere, where this decrease prevails, is called the *troposphere*; its upper limit is called the *tropopause*. In the tropics, where much insolation is received, the troposphere is thickest and reaches an elevation of about 16 000 metres. Near the poles, where the heating process is limited to the summer months and even then is relatively slow, the troposphere is thinner — it is generally not more than 8000 metres thick. Local and seasonal variations of these heights are quite large.

When cloudy conditions prevail, the terrestrial radiation heats up the lower atmosphere rapidly, because absorption of the heat is efficient and only the relatively thin layer of air below the clouds is affected. As soon as this lowest part of the atmosphere has reached the same temperature as the earth's surface, a balance is reached between the terrestrial radiation and re-radiation from the lower atmosphere downwards and temperatures will remain unchanged.

When no clouds are present, the terrestrial radiation heats a much thicker layer of the atmosphere. Terrestrial radiation is maintained by the temperature difference between the earth's surface and the lower atmosphere, resulting in a rapid decrease of temperatures at night at the lowest levels of the atmosphere.

Despite this supply of energy from the bottom, the atmosphere is not getting hotter in the long run. This is because the atmosphere also loses energy to outer space. This *outgoing radiation*, which is also of relatively long wavelengths, generally takes place from the top of the cloud cover. Only a relatively small proportion comes directly from the earth's surface, mainly when there are no clouds.

The intensity of the outgoing radiation from the earth and its atmosphere is proportional to the fourth power of the absolute temperature of the radiating surface. In this respect the conditions near the upper limits of the cloud cover are remarkably similar at all latitudes and during all seasons. In this region, close to the tropopause, the temperatures are usually round −50 °C. Therefore the average amount of outgoing radiation to outer space shows only very small variation with latitude and the annual values represent conditions very well (Figure 2.5). This diagram is based on figures for the northern hemisphere, but conditions are essentially the same for the southern half of the earth (Sasamori *et al.*, 1972).

The Radiation Balance

The graph of Figure 2.5 illustrates that in the annual balance between incoming solar radiation and outgoing radiation lost to outer space, the zones of surplus and deficit are of approximately the same size. The diagram indicates this, because its horizontal axis is divided proportionally to the surface of each latitudinal belt of 10 degrees. However, the location of these zones is not the same throughout the year. In January the surplus region extends from about 20° N to about 60° S; in July from 15° S to about 68° N (Simpson, 1929). The zone of surplus therefore moves with the apparent movement of the sun. The tropics belong to this zone all the time: they are the only region where continuously more energy is received than is lost to outer space. The polar areas, on the other hand, are always in the deficit zone: they lose more heat than they receive throughout the year.

There is no evidence, however, that the tropics are getting warmer, or the poles cooler. This is because there exists, within the atmosphere, a constant flow of energy from the tropics to the higher latitudes, compensating for the main latitudinal differences in the energy balance of the earth and its atmosphere. The form in which this energy is transported depends mainly on the nature of the earth's surface in the tropics (Sellers, 1965, pp. 103, 115). On a global scale, about half of it is carried as sensible heat, that is by the advection of warm air masses to colder regions. About 35 per cent is transported as latent heat, that is heat used when water is turned into water vapour. This latent heat is released when the water vapour condenses again in a cooler environment. The rest, or about 15 per cent of the energy, is carried by warm ocean currents which bring warm tropical water to the high latitudes. These ocean currents are, in turn, largely driven by the major wind systems of the atmosphere.

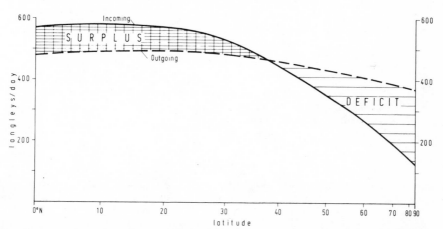

Figure 2.5   Annual radiation balance of the northern hemisphere (source: Houghton, 1954)

In whatever form the energy transfer takes place, its main source region is the tropics. It is here that a large part of the atmospheric circulation originates.

In conclusion, it can be established that there exists a continuous energy transfer, starting near the earth's surface in the tropics. This flow is directed upwards towards the higher parts of the troposphere and then horizontally away from the equator towards the mid-latitudes. The tropical circulation is completed by a flow towards the equator at relatively low levels, to compensate for the rising air masses near the equator. The driving force of this circulation is the surplus of solar energy over terrestrial energy losses to outer space, occurring near the equator.

References

Houghton, H. G., 1954, On the annual heat balance of the northern hemisphere, *Journal of Meteorology*, **11**, 7.

List, R. J., 1958, *Smithsonian Meteorological Tables*, 8th ed. Washington, D.C., Smithsonian Institution, 348 pages.

Sasamori, T., London, J. and Hoyt, D. V., 1972, Radiation budget of the southern hemisphere, *Meteorological Monographs*, **13**, 20.

Sellers, W. D., 1965, *Physical Climatology*, Chicago, University of Chicago Press, 272 pp.

Simpson, G. C., 1929, The distribution of terrestrial radiation, *Memoirs of the Royal Meteorological Society*, **3**, 53–78.

CHAPTER 3

Temperatures in the Tropics

When discussing temperatures, without further specifications, it is normally understood that reference is made to conditions near the earth's surface, where mankind lives. These temperatures are largely controlled by incoming and outgoing radiation, as described in the last chapter. However, a number of other factors also influence surface temperatures, and their distributions both over time and place are much more complicated than those of radiation, which are entirely controlled by the movements of the earth and which therefore vary regularly with latitude. Surface temperatures also show a correlation with latitude, but they show many deviations from the general pattern. Nevertheless, temperatures in the tropical climates have a number of characteristics in common, four of which are of major importance.

The first is the general and widespread *thermal uniformity*, which prevails both regarding the seasons and in relation to place. This uniformity is strongest near the equator and diminishes with increasing latitude: but is still considerable at the outer limits of the tropics in comparison with the higher latitudes.

The second major feature is the prevalence of the *diurnal cycle*. As seasonal variations are small, daily differences largely control the march of temperature in the tropics. Within the tropics, daily temperature ranges vary considerably, but they keep their importance everywhere.

The thermal uniformity over place is interrupted by the effects of *elevation*, which is the only factor that creates large temperature differences over short distances in the tropics. The influence of elevation has important practical consequences in the tropics.

A discussion of temperature conditions in the tropics is not complete without reference to their influences on organisms in general and on human beings in particular. These are indicated by the *physiological temperatures*, which show a number of characteristics typical to the tropics.

Thermal Uniformity

The absence of a cold season was mentioned in Chapter 1 as one of the distinguishing features of tropical climates. This means that temperature differences between the seasons are generally small. This *seasonal uniformity* is best illustrated by monthly temperature curves for some tropical stations (Figure 3.1). The stations chosen for representation are all in the southern

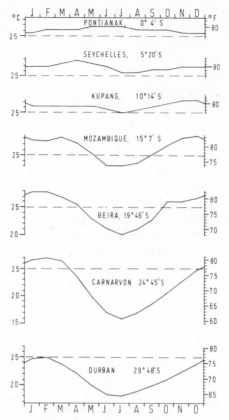

Figure 3.1 Monthly mean temperatures at selected stations.
(The location of these stations is indicated on Figure 3.2)

hemisphere, close to sea level and at marine locations, and at intervals of approximately 5 degrees latitude.

The main impression gained from these graphs is that the seasonal variation increases with latitude. The equatorial station shows almost no seasonal temperature differences. This is, of course, the direct consequence of the radiation conditions near the equator, where the incoming solar radiation is continuously strong (Figure 2.4). It should be noted, however, that the slight maxima of incoming radiation around the time of the equinoxes are not reflected in the temperature curve. This is typical for most equatorial stations and the reason is that these maxima are too small to have any effect on the surface temperatures, which are influenced much more by other factors, mainly cloudiness (Nieuwolt, 1968, p. 32). This is also why at Pontianak the highest monthly temperatures are recorded during the months of May to July, when the position of the sun is relatively low. During these months cloudiness is least, because the south monsoon comes mainly from the land side.

From about five degrees latitude onwards, however, the lower position of the sun and the slightly shorter days during June and July begin to have an effect on the surface temperatures. This is even the case at such an extremely marine location as the Seychelles. The effect of the seasonal reduction in incoming radiation becomes more prominent with further increases in latitude and it is during this season that there is a general drop of temperature with distance from the equator while during the opposite season of December and January the temperatures are approximately the same at all illustrated stations. There is, therefore, a gradual increase in the annual range of temperature with latitude, mainly due to cooler winters (Figure 3.1).

One important difference becomes evident when radiation and temperature curves are compared: the radiation curves are regular, controlled as they are by the movements of the earth only, but the temperature curves show many irregular variations. This is even true at the illustrated stations, which are at marine locations, where these irregularities are relatively small because the ocean waters keep their temperatures throughout the year at approximately the same level. The main reason for the irregular temperature variations are local factors.

The differences between stations are also far from regular, as is illustrated by the curves for Carnarvon and Durban. Carnarvon, on the western coast of Australia, is near a cold ocean current, while Durban, South Africa, is near the warm Agulhas Current. Consequently the winter temperatures at Carnarvon are lower than those at Durban despite its lower latitude. But the opposite relation prevails during the summer, when Carnarvon is frequently under the influence of continental air masses, while Durban has its rainy season, with much cloudiness (Figure 3.1).

The general increase of seasonal temperature differences with latitude can also be shown by the mean annual range (Figure 3.2). This map clearly indicates the strong effects of continentality and ocean currents. These exceed the influence of latitude over some of the oceans. Hence no clear zonal belts of annual temperature ranges are evident and this is the main reason why the mean annual range was rejected as a boundary of tropical climates (page 2).

Another seasonal temperature characteristic is found at many tropical stations from about 10 degrees latitude onwards. It is the so-called 'Ganges-type' of temperature regime, in which the temperatures are higher in the spring than in the summer (Köppen, 1931). This regime is often associated with monsoonal wind systems because it is particularly well developed in monsoon Asia; but it also occurs at some continental stations which have no monsoons (Figure 3.3). At marine stations it is rare. This feature is largely caused by cloudiness. The main rainy season is in the summer and heavy cloudiness prevails, so that temperatures remain relatively low. But in the spring and, to a lesser extent, again after the rains, temperatures go up very rapidly during the sunny days and this effect more than compensates for the slightly shorter days and lower positions of the sun compared to mid-summer. It is clear that this type does not occur near the equator and at marine locations, where clouds are frequent even during the drier parts of the year.

18

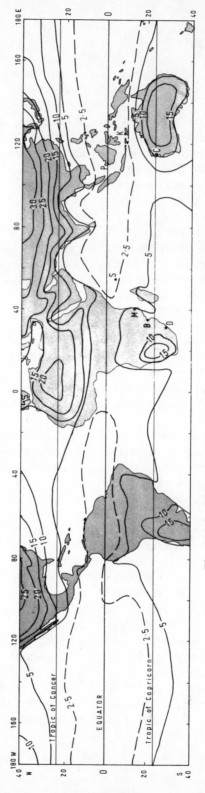

Figure 3.2 Mean annual range of temperature at sea level (degrees C).
(Letters indicate stations used in Figure 3.1)

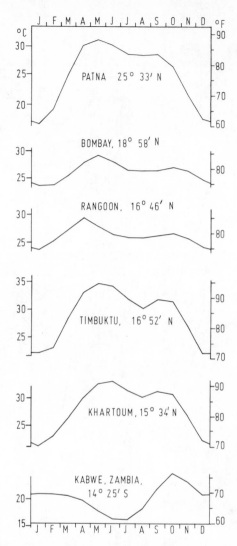

Figure 3.3　Monthly mean temperatures at some monsoonal and continental stations in the tropics.
(The location of these stations is indicated on Figure 3.4, July map)

Despite these local variations, temperature differences with place are generally quite small in the low latitudes. This is illustrated by the sea level isotherms for January and July (Figure 3.4). Thermal gradients are very small near the equator and over the oceans, as shown by the large distances between the isotherms. Towards the outer limits of the tropics thermal differences with place increase rapidly, especially over the continental areas and during the respective winter months.

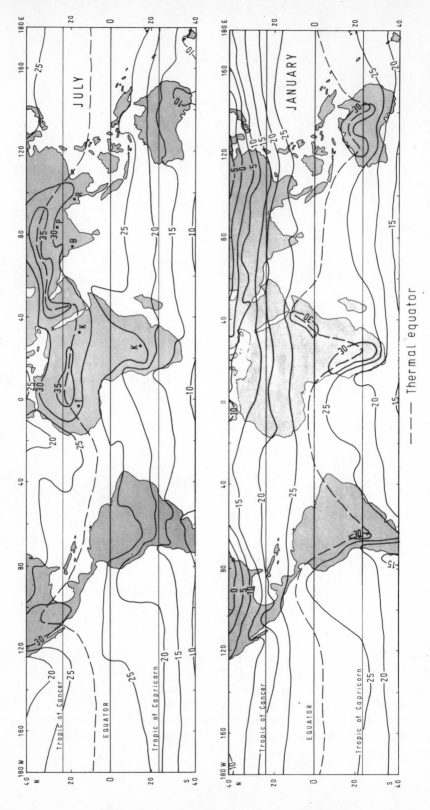

Figure 3.4   Sea level temperatures in July and January (°C). The position of the thermal equator (temperatures reduced to sea level).
(Letters indicate stations used in Figure 3.3)

- - - Thermal equator

The main reason for this thermal uniformity with place in the tropics is the small differences in the amount of net radiation received in various parts of the low latitudes: between 0 and 10 degrees of latitude the curves are rather close throughout the year (Figure 2.4). Large temperature differences are therefore unlikely to develop. A secondary factor is the high proportion of ocean surfaces in the tropics. Oceans act as enormous heat storage reservoirs. Air masses, moving over the oceans, pick up some of this heat in the form of sensible heat and latent heat of vaporization. When these air masses move into continental areas, they gradually release this heat when they are forced to rise, either by orographic lifting or by convection. This exchange of heat between oceans and continents is greatly facilitated in the tropics, where there are no large continents. It prevents the formation of really cold air masses during the winter.

The maps (Figure 3.4) also show where the highest sea level temperatures occur. These are indicated by the 'thermal equator' which is not an isotherm since it connects maxima of different intensity. Over the oceans the thermal equator remains close to the geographic equator throughout the year. It follows the movements of the overhead position of the sun with a delay of about one month, but the extent of these periodic movements in a meridional direction is limited to a few degrees of latitude only. Near cold ocean currents, such as the Humboldt and Benguela Currents, the position of the thermal equator is continuously on the northern hemisphere.

Over continents, however, the thermal equator makes large seasonal movements, amounting to as much as 40 degrees of latitude. During the summers the combined effects of long summer days and a high position of the sun produce very high temperatures at latitudes of about 20–25 degrees. These effects are particularly pronounced over highlands, such as in Central Asia, Iran, Ethiopia and Mexico. Here the intense incoming radiation at high elevations causes high daytime temperatures, which are not strongly reduced during the relatively short summer nights. When these upper level temperatures are reduced to sea level, they stand out as extremely high and seem rather unrealistic. The high January temperatures in Ethiopia during January, which is the winter season, are largely due to the lack of clouds in that period.

The thermal equator is closest to the geographic equator shortly after the equinoxes, that is during April and October. However, its meridional movement is rather rapid at that time and it remains near the equator for only a short period.

Diurnal Temperature Variations

In many parts of the tropics the seasonal temperature differences are so small that they are of little consequence. In these areas, mainly near the equator and over the tropical oceans, temperature conditions are almost entirely dominated by diurnal variations. These are generally expressed as the *mean diurnal range*, indicating the difference between daily maximum and minimum temperatures.

22

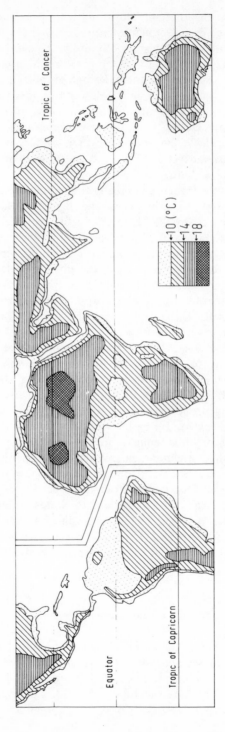

Figure 3.5 Mean daily range of temperature (after Paffen, 1966)

The importance of this indicator as a climatic characteristic of the tropics was mentioned in Chapter 1 (page 2).

Unfortunately, over large parts of the tropics, and especially over the oceans, no sufficient data are available to compute the mean diurnal range of temperature. This aspect of climate has therefore been studied mainly in the land areas of the tropics (Paffen, 1966, 1967). Its distribution pattern is rather complicated (Figure 3.5). This is caused by the strong effect of local factors, which often control temperature conditions temporarily. As the mean diurnal range expresses temperature variations over rather short periods, of the order of about twelve hours, these local factors become a major influence where they are effective throughout the year. The main one is *continentality*. All marine locations display very small diurnal ranges, which increase rapidly with distance from the sea. Large inland water bodies, such as Lake Victoria, Lake Nyasa, or large swampy areas in the Congo and Amazon basins, produce local reductions in the mean diurnal range. Highest values are reached in the dry continental areas.

A second factor which controls the mean diurnal range is *elevation*. However, the effects of this factor differ according to the prevailing topography. In extensive highlands the diurnal range generally increases with elevation, as both incoming and outgoing radiation become more intense at higher levels. This effect is particularly strong over dry highlands, where radiation is strong because of little cloudiness (Paffen, 1966, pp. 261–263). Air drainage, which reduces temperatures at night in basins and valleys, helps to increase the diurnal temperature range locally.

On isolated mountain tops the effect of elevation on the diurnal range of temperature is quite different: because the exchange with the free atmosphere is facilitated and the influence of the earth's surface reduced, the diurnal temperature range generally decreases with higher altitude. This effect is intensified by clouds, it is particularly clear at lower elevations near the equator, where cloudiness increases with height, so that direct radiation is greatly reduced (Nieuwolt, 1969, pp. 44–45).

The third factor which controls the diurnal temperature range is *cloudiness*. This factor has widespread effects, as illustrated on the world map (Figure 3.5). The diurnal range is generally low near the equator and increases generally towards the dry tropics at about 15–25 degrees latitude, where it reaches maximum values. This is largely caused by differences in cloudiness. A heavy cloud cover strongly reduces both incoming and outgoing radiation and thereby the diurnal temperature differences. Locally, clouds frequently reduce the diurnal range in mountainous areas, where clouds develop during the day as the result of convection.

In respect of diurnal temperature variations there exists a qualitative difference between the tropics and the mid-latitudes, which is of great significance. In the tropics, where the duration of day and night remains about the same throughout the year, the diurnal cycle shows generally only small seasonal variations. Those which occur are usually related to changes in

24

cloudiness (Nieuwolt, 1968, p. 38). Irregular changes in the diurnal march of temperature in the tropics are also generally small. They are mainly caused by disturbances, which are predominantly of the thunderstorm type and therefore of short duration (Nieuwolt, 1966). These irregular changes have, therefore, little effect on the mean diurnal range of temperature. The diurnal temperature cycle in the tropics is consequently rather regular and the day-to-day variations are generally quite small.

In the higher latitudes, however, the short-term temperature conditions are largely controlled by frequent fronts and depressions, which bring rapid changes in air-masses of strongly contrasting temperatures. The diurnal march of temperature is therefore rather irregular. It also varies strongly with the seasons, as the duration of day and night and the amount of incoming radiation differ during the course of the year.

It is this regularity of the diurnal temperature regime in the tropics which has caused the frequent use of terms such as 'monotony', 'uniformity' and even 'weather synonymous with climate' in many descriptions of low-latitude

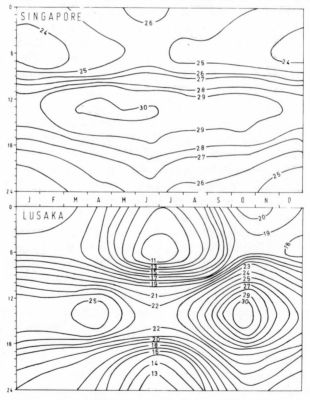

Figure 3.6   Thermoisopleths for Singapore (1°21′ N, 8 m above sea level) and Lusaka (15°25′ S, 1154 m a.s.l.) (degrees C).
(Based on records over 1951–1960 (Singapore) and 1941–1951 (Lusaka))

climates (Critchfield, 1966, p. 178; Koeppe and de Long, 1958; Trewartha, 1954, p. 242). These terms completely disregard the temperature variations from day to day which occur almost everywhere in the tropics. Though small when expressed in numbers, they are of great importance to the inhabitants of the tropics, who are very sensitive to small temperature differences (Nieuwolt, 1968, p. 36.).

To illustrate the combined effects of both the seasonal and the diurnal variations of temperature, thermoisopleths are the best method (Figure 3.6). The diagram for Singapore shows a typical pattern at an equatorial station: temperature conditions are completely dominated by diurnal variations. Though the mean diurnal range is rather small at this marine location, the diurnal cycle is still more important than the seasonal differences. Similar conditions prevail at most equatorial stations.

The Lusaka diagram illustrates a more balanced situation, with the diurnal and seasonal variations of about equal importance. These conditions are characteristic of areas near the outer limits of the tropics. However, local deviations from this type are generally larger and more common than near the equator. Lusaka shows the spring maximum in October, and a slight increase in temperature after the rains, in March and April, which is rather typical for a continental tropical station at this latitude. The diurnal range is considerable throughout the year, the result of both elevation and Lusaka's continental location. Marine stations at this latitude would show these features to a much smaller extent.

Neither Lusaka nor Singapore represent extreme situations, but the two diagrams illustrate the extent of differences that exist between the various tropical climates in respect of temperature regimes.

The Effects of Elevation

The general uniformity of temperature with place which prevails in the tropics is interrupted by one major factor: elevation. It can cause large temperature differences over short distances and this is why the world maps used in this chapter all show temperatures reduced to sea level. The main purpose of this reduction is to exclude the effects of elevation, thereby keeping the illustrated distribution patterns relatively simple.

Temperatures always go down with elevation, but the rate of the decrease, the lapse rate, is far from uniform (Lautensach and Bögel, 1956). It not only varies with cloudiness and, therefore, in many areas with the seasons, and between day and night, but also depends on the prevailing topography of the highlands. As was discussed in the description of the mean diurnal range, the main difference in this respect exists between extensive highlands and isolated mountain tops.

Because the lapse rate is so variable, a general addition of 5 °C or 6 °C for every thousand metres that a station is above sea level can introduce considerable errors. Where the lapse rate is generally small, the resulting sea level temperatures can be unrealistically high, as is the case near some tropical highland regions (p. 20).

The main controlling factor on the lapse rate in mountainous regions is *cloudiness*. Under cloudy conditions, when the effects of radiation are greatly reduced, the lapse rate tends to become similar to that of the free atmosphere, which is usually around 5 °C/1000 m. This value will prevail during both the day and the night.

But when the sky is clear, diurnal differences in the lapse rate are large. During the day the absorption of solar radiation will increase the temperatures at all levels, but this effect is especially strong at higher altitudes, where insolation is more intense as there is less water vapour and dust in the atmosphere. The lapse rate will consequently be rather small. But, at night, the heat loss by terrestrial radiation will be strong, and again most effective at higher elevations, so that the lapse rate will grow rapidly during the night. These diurnal differences are particularly large over tropical highlands, where day and night are of about equal duration and where insolation is strong.

Still, the effects of cloudiness on the lapse rate cannot easily be generalized over large areas, because of its variability. In many cases clouds will develop over the highlands during the day, mainly as the result of local convection and orographic lifting, while the lowlands remain clear. There can also be large differences between windward and leeward side of a mountain. It can therefore only be stated that the lapse rate in the tropics is generally larger in the dry seasons than in rainy periods, and therefore higher in the dry tropics than near the equator (Lautensach and Bögel, 1956, pp. 277, 280).

The extent of the local variations of the lapse rate can be illustrated by conditions in Kenya, where the annual mean in the eastern highlands is about 7 °C/1000 m, while in the western parts it is around 6 °C/100 m When annual extreme temperatures are considered, the differences are even larger, as the lapse rates vary between 3·1 and 9 °C/1000 m (Kenworthy, 1966).

Most tropical settlements at higher elevations are, of course, not on isolated mountain tops, but on extensive highland regions. Therefore the lower temperatures at higher levels are usually accompanied by large diurnal ranges of temperature (Trewartha, 1954, p. 370). The mean annual range, on the other hand, seems largely unaffected by elevation; it shows only very small and irregular differences between lowland and highland stations in Malaya (Nieuwolt, 1969, pp. 39–40).

The importance of cool highland areas in the tropics depends largely on two functions: they bring relief from the oppressive heat of the tropical lowlands and they offer special agricultural possibilities in a region of otherwise rather uniform temperature conditions. Highland stations based primarily on the first function exist in many parts of the tropics. Bandoeng (717 m above sea level) in Java, near Djakarta, Simla (2200 m) not far from New Delhi and Fraser's Hills (1300 m) in Malaya, close to Kuala Lumpur, are examples of settlements which provide a temporary cool environment for government employees from the capital cities. And when the British made Nairobi (1660 m) the capital of Kenya, the pleasant climatic conditions of the city were certainly a favourable factor and made it preferable to Mombasa, the harbour city on the Indian Ocean, with its hot and humid climate. Similarly, Mexico City (2260 m) is favoured with a much cooler climate than the coastal lowlands of Mexico. However, not only more pleasant climatic conditions, but also the absence of some tropical diseases in the highlands made the choice of these sites almost a necessity in times when most administrators came from the mid-latitudes.

The agricultural possibilities of the tropical highlands are caused by the different temperature conditions at various levels: the optimum temperatures for many crops or other forms of agriculture can easily be found by choosing the correct elevation. Moreover, as the temperatures exhibit only small seasonal variations, production can be sustained throughout the year and risks related to the temperature factor are very small. Added advantages of the tropical highlands are sufficient rainfall in most of them and thermal comfort for the inhabitants. The latter feature was particularly important when plantations were established, as these were frequently run by managers and technicians from the mid-latitudes.

In relation to prevailing temperatures, a number of altitudinal zones of various agricultural activities can be recognized in most tropical highlands. It must be emphasized that the boundary levels indicated below are only general indications of zones of gradual transitions. These limits vary locally, especially in relation to annual total and seasonal distribution of rainfall (Trewartha, 1954, pp. 372–373; Troll, 1959). Other variations are due to topography. Factors such as slope-steepness and aspect in relation to rain-bringing winds and sunshine create microclimatic conditions which may differ considerably from the general climate. Moreover, altitudinal limits move downwards with increasing distance from the equator until the lower zone disappears completely near the outer boundaries of the tropics. With these reservations in mind, the following zonal belts can be recognized near the equator:

(1) *the lowlands.* The cooling effects of elevation are generally not strong enough to have any influence on the agricultural activities up to approximately 500 m above sea level. In Latin America this zone is called 'tierra caliente'; it is characterized by annual mean temperatures around 24–27 °C. and the normal tropical lowland crops are cultivated. Their choice and distribution is largely governed by rainfall conditions rather than temperature. In the upper levels of this zone some of the crops of the next zone, notably coffee, are sometimes grown where rainfall conditions are favourable.

(2) *The temperate zone.* From about 500 m to approximately 2000 m above sea level the temperatures are considerably lower than in the lowlands. The annual mean temperature in this 'tierra templada' varies between 16 °C and 24 °C. Under these conditions many different crops can be cultivated. Some lowland crops, such as sugar cane, rice and maize do very well in this zone, while cotton, bananas and pineapples are mainly limited to the lower parts. Tropical highland crops, such as tea and coffee, are widely distributed. In addition a number of mid-latitude crops can be grown successfully: beans, citrus fruits and vegetables in

the wetter parts, grain crops, such as wheat and barley, in the drier regions. This is therefore a highly productive zone and the choice of crop combinations depends mainly on microclimatic conditions and the seasonal distribution of rainfall.

(3) *The cold zone,* or 'tierra fría', is situated between about 2000 m and 3000 m above sea level. Here, temperatures are greatly reduced in comparison with the lowlands and the annual mean is between 12 °C and 18 °C. In this zone the tropical highland crops prevail. Tea, especially, reaches high levels where rainfall is adequate and no night frosts occur. These conditions depend largely on local topography. Another important crop of this zone is pyrethrum, which is mainly found in the wetter areas. In the drier parts wheat, barley and mid-latitude fruits are widely grown. This zone is largely above the upper level where rainfall increases with elevation, and rainfall conditions are therefore more critical than in most other parts of the tropical highlands (pp. 107).

(4) The next zone, between about 3000 m and 4000 m above sea level, is often called the *paramó belt,* named after a low grass and bush formation which prevails at these heights in the Andes mountains. The annual mean temperature varies between 6 °C and 12 °C. Depending on the amount of rainfall normally received, agricultural production is concentrated on dairy cattle, which can be fed on the alpine meadows and pastures, or on mid-latitude crops such as barley and potatoes, which can withstand low temperatures well. However, many parts of this zone are not cultivated and in the wetter parts, where forests form the natural vegetation cover, these have largely remained uncleared.

(5) The *frost zone,* above 4000 m approximately, where snow and ice prevail. In this 'tierra helada' no agricultural land use is possible.

The belts described above refer to tropical highlands and it is possible to observe a similar zonation in the mid-latitude mountains, where equivalent boundaries would occur at lower levels (Trewartha, 1954, p. 372). However, there is one important difference: outside the tropics seasonal temperature variations are generally large and they control the agricultural production, limiting most activities at higher levels to the summer only. They also cause seasonal movements, such as the transhumance of the Mediterranean regions, in search of better agricultural possibilities. In the tropics, seasonal temperature variations are small. If seasonal differences are important, they are usually a matter of rainfall rather than temperature, and therefore not related to specific elevations.

Another difference between tropical and mid-latitude mountains is that the zonal belts outside the tropics are generally much narrower. Though the lower zones of the tropics are absent, the remaining belts are telescoped into a much smaller range of elevation, because the zone of snow and ice starts at a much lower level compared to tropical mountains.

Physiological Temperature

The temperature, as experienced by living organisms in general, and by human beings in particular, depends mainly on the rate of heat loss from the body. The human body is kept at a constant temperature of 36·7 °C, and defence mechanisms prevent excessive loss of heat or too much heat absorption. Under warm conditions, heat disposal takes place mainly from the skin and the lungs but, if this is not sufficient, it is greatly increased by the evaporation of body fluids in the form of perspiration from the skin. Physiological temperature therefore does not depend only on the temperature of the air, but also on the efficiency and speed of evaporation, which is controlled by a number of other factors.

The most important of these factors is the humidity of the air. When the air is humid, the evaporation of perspiration is limited and a feeling of oppressiveness is created. The combination of high temperatures and high humidities is sometimes indicated as the sultriness of the air (Blüthgen, 1966). Dry air, on the other hand, allows a rapid cooling of the skin as much latent heat is used for the evaporation of the perspiration and high air temperatures are endured much better with low humidities.

A second factor which influences physiological temperature is the circulation of air around the body. Where air is stagnant, the layer of air around the body not only heats up to approximately body temperature, but is also soon saturated with water vapour, thus restricting further cooling of the skin by evaporation. When sufficient ventilation is available, the constant replacement of air around the body prevents this situation arising.

A third factor is direct exposure to solar radiation, In the sunshine the skin absorbs a great deal of heat although the actual air temperature is possibly the same as in the shade.

However, the main factor controlling physiological temperature remains the temperature of the air. The other factors only contribute to differences in the sensation the air temperature creates in living organisms. This sensation depends on individual characteristics, such as general build and weight, on type of clothing worn and on the physical activities carried out. It is also influenced by psychological factors and adjustment to the prevailing climatic conditions. It cannot therefore be measured or calculated directly, but is usually expressed as the percentage of a number of test-persons feeling thermal stress under carefully controlled laboratory conditions.

When quantitative relations between environmental conditions and the reactions of human beings are investigated, the emphasis is usually only put on the effect of temperature and humidity of the air. These two factors are easily controlled and measured and it is generally accepted that they are the two most important factors controlling the physiological temperature. The two other factors, ventilation and exposure to sunshine, can easily be changed in normal living conditions. The electric ventilator is by far the most widespread and cheapest instrument for alleviating thermal stress in hot climates, and simpler types of ventilators have been in use for thousands of years. Protection against exposure to sunshine during outdoor activities is usually provided by sunshades, protective clothing or trees.

When physiological temperature is expressed in relation to temperature and humidity of the air, a large number of indices can be used (Bedford, 1940;

Critchfield, 1966, pp. 347–349; Smith, 1955; Stephenson, 1963; Terjung, 1966; Webb, 1959, 1960). Probably the most widely used is the Temperature–Humidity–Index (THI), sometimes called the Discomfort Index (Thom, 1958; US Weather Bureau, 1959). This index expresses thermal stress in degrees of temperature. It indicates the temperature which, combined with a relative humidity of 100 per cent, would create the same reactions by test persons as the existing combination of temperature and humidity. It was originally devised for use with the Fahrenheit scale, but it can easily be converted to degrees Celsius. The simplest formulae to express the index in degrees Celsius are:

$$THI = 0{\cdot}8t + \frac{RH \times t}{500} \quad \text{or}$$

$$THI = 0{\cdot}55t + 0{\cdot}2d + 5{\cdot}3$$

in which t is the air temperature, RH is the relative humidity (%) and d the dewpoint temperature.

Generally, a THI of around 21 °C made most people feel comfortable. At values of about 24 °C half of the people interrogated experienced some form of thermal stress; and when the RHI reached 26 °C almost all felt uncomfortably hot. With further increases of the index the efficiency of workers deteriorated rapidly (US Weather Bureau, 1959). These figures are based on experiments in the mid-latitudes. It is quite possible that in the tropics, where exposure to thermal stress occurs more frequently, higher values of the THI are better tolerated. This is a matter not only of acclimatization but also of habits of nutrition, clothing and general physical activities. The latter are usually reduced in warm climates and the speed of everyday life in the tropics is therefore normally lower than in the mid-latitudes.

When the effects of ventilation on physiological temperature are considered, they are usually called 'effective temperatures' and computed from nomograms (Ashrae, 1967; Bedford, 1940; Miller, 1964; Smith, 1955, p. 9). When the THI is used, the effects of air movement can be included by reducing the index by the factor: $1{:}4v$, in which v is the wind speed in knots, and the THI is expressed in degrees Celsius (Webb, 1959, 1960). However, the influence of the air movements can only be considered where reliable data about wind speeds are available. Owing to the gustiness of the wind and the very strong effects of local conditions, wind figures are often of poor quality and it is certain that wind speeds at airfields do not correspond at all with conditions in cities. Data for different stations are often not comparable because of dissimilar instrumentation, exposure and observation methods. As air ventilation can easily be controlled or generated in most human environments, its influence on physiological temperature conditions is frequently ignored.

The diurnal and seasonal variations of the THI (excluding the effects of winds) at two representative tropical stations are illustrated in the form of isopleth diagrams (Figure 3.7). A comparison with the graphs for temperature

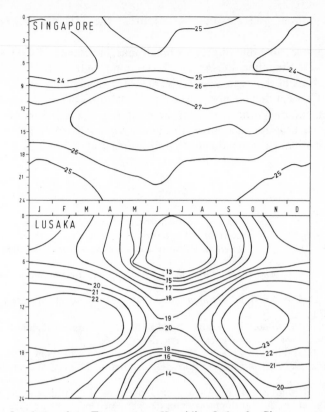

Figure 3.7   Isopleths of the Temperature–Humidity–Index for Singapore and Lusaka.
(Based on records over 1951–1960 (Singapore) and 1966–1970 (Lusaka))

at the same stations shows that the general patterns of the curves are very similar
(Figure 3.6). The main difference between the two sets of diagrams is that the
diurnal variations of the THI are much smaller than those of temperature alone.
This is caused by the relative humidity of the air, which is generally much higher
at night than during the day. It therefore increases the THI more during the
night than in the daytime, thereby reducing its diurnal variation.

At Singapore the seasonal variation of the THI is similar to that of the air
temperature. This is caused by the uniformity of the relative humidity, which
shows an annual range of only a few per cent (Nieuwolt, 1969, p. 132). Such
conditions are rather typical for equatorial regions where no dry season occurs.
Despite its reduction by the diurnal differences of the relative humidity, the
diurnal cycle of the physiological temperature prevails in Singapore, confirming
the old adage that the nights of the tropics are their winter.

At Lusaka, the seasonal variations of the THI are smaller than those of air
temperature. This is caused by the low relative humidity during the hottest
months of October and November, which reduces the THI during that period.

Here, the diurnal and seasonal variations of the THI are of about equal importance.

The diagrams show that oppressive conditions, indicated by a THI of 26 °C or over, prevail in Singapore during the afternoon throughout the year. At Lusaka a thermal stress of this type is very rare, but rather cool nights, with THI values below 18 °C, occur during most of the year. Conditions of this pattern are rather typical for stations near the outer limit of the tropics. However, diurnal variations are usually much larger in the dry tropics. The graphs illustrate the importance of elevation in reducing the oppressive heat of the tropical lowlands.

References

ASHRAE, 1967, *Handbook of Fundamentals*, American Society of Heating, Refrigerating and Air-conditioning Engineers, New York, pp. 117–118.

Bedford, T., 1940, *Environmental Warmth and its Measurement*, Medical Research Council, London, War Mem. No. 17, p. 24.

Blüthgen, J., 1966, *Allgemeine Klimageographie*, 2nd ed., Berlin, Walter de Gruyter, pp. 120–122.

Critchfield, H. J., 1966, *General Climatology*, 2nd ed., Englewood Cliffs, N.J., Prentice-Hall, 420 pp.

Kenworthy, J. M., 1966, Temperature conditions in the tropical highland climates of East Africa, *East African Geogr. Review*, No. 4, 4–6.

Koeppe, C. E. and de Long, G. C., 1958, *Weather and Climate*, New York, p. 201.

Köppen, W., 1931, *Grundrisz der Klimakunde*, Berlin, Leipzig, Walter de Gruyter, 388 pp.

Lautensach, H. and Bögel, R., 1956, Der Jahresgang des mittleren geographischen Höhengradienten der Lufttemperatur in den verschiedenen Klimagebieten der Erde, *Erdkunde*, **10**, 270–282.

Miller, A. A., 1964, *The Skin of the Earth*, London, Methuen, pp. 174–175.

Nieuwolt, S., 1966, The urban microclimate of Singapore, *Journal of Tropical Geography*, **22**, 31–34.

Nieuwolt, S., 1968, Uniformity and variation in an equatorial climate, *Journal of Tropical Geography,* **27**, 23–39.

Nieuwolt, S., 1969, *Klimageographie der malaiischen Halbinsel*, Mainzer Geographische Studien, No. 2, Mainz, 152 pp.

Paffen, K., 1966, Die täglichen Temperaturschwankungen als geographisches Klima-charakteristikum, *Erdkunde*, **20**, 252–265.

Paffen, K., 1967, Das Verhältnis der Tages- zur jahreszeitlichen Temperatur-schwankung, *Erdkunde*, **21**, 94–111.

Smith, F. E., 1955, *Indices of Heat Stress*, Medical Research Council, London, Memorandum No. 29, pp. 8, 9.

Stephenson, P. M., 1963, An index of comfort for Singapore, *Meteorological Magazine*, **92**, 338–345.

Terjung, W. H., 1966, Physiologic climates of the coterminous United States: a bioclimatic classification based on man, *Annals, Association of American Geographers*, **56**, 141–179.

Thom, E. C., 1958, Cooling degree days, *Air Conditioning, Heating and Ventilating*, **22**, 65–72.

Trewartha, G. T., 1954, *An Introduction to Climate*, 3rd ed., New York, McGraw-Hill, 402 pp.

Troll, C., 1959, *Die tropischen Gebirge. Ihre dreidimensionale klimatische und pflanzengeographische Gliederung*, Bonner Geogr. Abhandl., Vol. 25, Bonn, 112 pp.

US Weather Bureau, 1959, *Notes on Temperature-Humidity Index*, L.S.5922, Washington, D.C., 4 pp.

Webb, C. G., 1959, An analysis of some observations of thermal comfort in an equatorial climate, *British Journal of Industrial Medicine*, **16**, 297–310.

Webb, C. G., 1960, Thermal discomfort in an equatorial climate, *Journal of the Institute of heating and ventilation engineers*, **7**, 1–8.

CHAPTER 4

The General Circulation of the Tropical Atmosphere

When trying to describe and explain the general pattern of air movements in the tropical atmosphere, one is faced with three groups of difficulties, related to the nature of the subject, the lack of sufficient and reliable information and the frequent deviations from the general circulation pattern.

Describing the general circulation of the atmosphere, one considers air movements, changeable both in direction and velocity over largely different periods of time, in a three-dimensional medium. The whole complicated structure can be illustrated only by a combination of maps and cross-sections, drawn for different seasons; but even so it is difficult to visualize in its entirety the general circulation and its main changes with the seasons.

In this respect it must be stated that meteorologists, when illustrating conditions in the atmosphere, usually exaggerate the vertical dimensions and movements, without actually stating so, thereby creating an impression of great depth. However, the troposphere, in which the general circulation takes place, is extremely shallow: its maximum depth, near the equator, is around 20 km, while its horizontal dimensions, in any direction, are measured in thousands of kilometres. Encircling the globe, horizontal distances are practically infinite within the troposphere. Because of this shallow depth, horizontal air movements prevail in the troposphere and the masses of air transported parallel to the earth's surface are many times larger than those carried by vertical air currents. Nevertheless, the latter are of great importance, because they transport heat and moisture away from the lowest levels and also because they profoundly change the stability conditions of the air masses involved. This is, of course, the main reason why vertical movements are emphasized in most meteorological literature.

A second group of problems arises from the widespread lack of upper air observations over many parts of the tropics. Upper air observations are expensive to obtain and the meteorological services in many tropical countries are therefore forced to concentrate their efforts, kept rather modest by economic conditions, to surface data, which are of limited use in the study of the general circulation. Only a few areas, where intensive studies of strong tropical disturbances have been carried out, such as the Caribbean region and the South China Sea, have dense networks of upper-air observation points, but one cannot extrapolate findings in these regions to other parts of the tropics.

The third group of difficulties encountered is related to this lack of information: it has been established that many deviations from the general pattern of the circulation in the tropics occur. These variations are effective over highly different areas and times: some affect whole continents or ocean basins, others are purely local, limited to coastlines, islands or mountain ranges; some are seasonal in character, others are diurnal in occurrence and still others come and go at irregular intervals. As upper-air observations over large parts of the tropics are scarce, it is often difficult to estimate the size, importance and duration of these deviations accurately. This makes it hard to recognize a basic general circulation of the tropical atmosphere. 'No single meridional cross-section can be considered as representative of the tropical circulation' (Riehl, 1954, p. 24). Nevertheless, especially over the oceans, a rather persistent system can be recognized, which is described in this chapter. This description is gathered together from many sources, often with contradictory evidence and conclusions, which were carefully assessed and compared, resulting in a picture of the whole circulation without inconsistencies. It is a realistic picture, as it is valid over large areas, and the day to day variations from the mean pattern are generally much smaller than in the higher latitudes. The most important regular deviations from the general circulation pattern will be the subject of Chapter 5, and the smaller disturbances, superimposed on the large-scale movements of air, will be dealt with in Chapter 6.

Generally, problems of short-term weather forecasting in the tropics are much less urgent than in the extra-tropical areas, where strong disturbances may change the weather picture dramatically at short notice. Therefore, meteorological research in the tropics has been rather limited until fairly recently. It was widely believed that the general circulation of the tropical troposphere was rather simple and regular, and not a worthwhile subject of study. It was only during and after the Second World War, with its many activities over tropical areas, that it was discovered that many new observations did not agree at all with this view. It was also realized, at approximately the same time, that many occurrences in the tropical atmosphere have close relations to meteorological conditions in the mid-latitudes.

The tropics, as delimited in Chapter 1, cover more than 40 per cent of the earth's surface, but the tropical circulation often extends beyond these boundaries. Moreover, the thickness of the troposphere is higher over the low latitudes than further away from the equator. The tropical circulation therefore involves the main bulk of the troposphere and the extra-tropical parts can be considered as two relatively minor and shallow appendices of it (Beckinsale, 1957).

The importance of the tropical circulation is therefore related not only to its control of the main characteristics of the tropical climates, but also to its influence on conditions outside the tropics.

The Hadley cell

One of the main functions of the tropical circulation is to dissipate to other parts of the globe the surplus heat energy supplied to the low latitudes by the intensive

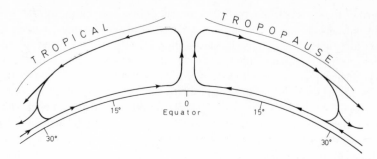

Figure 4.1   Meridional cross-section of the tropical troposphere (after Palmén, 1951). Vertical exaggeration: about 100 times

solar radiation (p. 13). The constantly maintained difference in temperature between the tropics and the higher latitudes is the main driving force of the meridional air currents of the tropical circulation. In a simplified and generalized model, consisting of a closed thermally-driven circulation over each hemisphere between the equator and about 30 degrees latitude, this is illustrated by the Hadley cell (Hadley, 1735).

According to this model, the surplus heat energy is transformed into kinetic energy near the equator (Figure 4.1). A large proportion is also used for evaporation and carried with the moving air masses as latent heat, which is released when condensation occurs. The heated air masses rise to high levels and, on their way, large-scale condensation takes place, resulting in further warming. The air masses are then transported to higher latitudes by upper air currents, generally called the anti-trades, which gradually slow down at about 20–30 degrees latitude. A subsiding movement prevails, taking air back to the earth's surface and resulting in areas of relatively high surface pressure, the subtropical highs. The circulation is closed by a massive air stream moving at low levels towards the equator, the trade winds. The two trade wind systems converge at an area of low pressure, called the equatorial trough or Inter Tropical Convergence Zone.

This model has many defects. The main one is that it fails to show both seasonal variations and differences with longitude, as illustrated on maps of the surface positions of its main components in January and July (Figure 4.2). In its basic form, the Hadley cell only exists over the Pacific and Atlantic Oceans. At most other longitudes it is strongly distorted and displaced, but it must be emphasized here that the maps show conditions during the solstices; in the periods around the equinoxes the general situation is much more similar to the Hadley model at most longitudes.

Another important defect of the simplified model is that it assumes a single heat source near the equator, while in reality there are often a number of separate areas of surplus heat, both over continents and oceans, so that the centre of the circulation varies in width and intensity, as well as in latitudinal position.

Figure 4.2 Mean sea level pressure and predominant surface winds in January and July. Dashed line: mean position of the I.T.C.Z.; dotted line: secondary convergence areas

The model also ignores inter-hemispheric exchanges of heat and it is now known that these occur frequently, mainly from the southern to the northern hemisphere (Alaka, 1964; Asnani, 1967). Air currents in the middle troposphere are also neglected in the model. Sometimes large meridional troughs prevail over many degrees of latitude in the tropics and along these troughs a strong poleward flow of heat takes place (Riehl, 1950; Rossby, 1947).

Despite these imperfections, the Hadley model presents a reasonably accurate picture of the main air movements in meridional directions over the tropical oceans.

Most explanations of the basic Hadley cell circulation emphasize either the temperature conditions or the movements of the air themselves as the main factor responsible for its origin and continuation. *Thermal theories*, in their most basic form, relate the Hadley cell directly to the large latitudinal differences in temperature resulting from radiation conditions. High temperatures near the earth's surface cause low pressure and rising air masses; lower temperatures conversely produce high pressure near the surface and subsiding air movements. A pressure gradient between zones at 20–30 degrees of latitude and the equatorial areas therefore prevails frequently and the trade winds are mainly the consequence of these pressure differences.

Dynamic explanations of the Hadley cell emphasize the self-reinforcing nature of some of the main air movements. Instability of equatorial air masses leads to rising movements, conservation of angular momentum to subsiding movements in the subtropical highs. In this way of thinking, pressure differences are largely the result of air movements, and they merely help to maintain the circulation.

These two groups of explanations are generally complementary rather than contradictory. Both contain valid generalizations and are supported by some, and contradicted by other, observations. It is, at present, still impossible to evaluate the relative importance of the various factors indicated as the main causes of the Hadley cell circulation.

The central zone of low pressure and convergent air masses

At the centre of the tropical circulation is a wide belt, characterized by relatively low surface pressure, rising air movements and convergence of air masses (Figures 4.1, 4.2). This zone is known by various names, depending on which of its main characteristics is considered the most important. When the low pressure is emphasized, it is usually called the 'equatorial trough', 'heat low' etc., but when the convergence of air masses is considered more relevant, it is called the 'Inter Tropical Convergence Zone', 'trade confluence' or 'near-equatorial convergence'. At times the term 'intertropical front' was used for this zone, by analogy with the mid-latitude fronts between convergent air masses, but this name is now largely out of use since it was realized that the convergence zone near the equator rarely displays the characteristics of a mid-latitude front, because temperature differences between converging air masses are usually very small. Finally, a more neutral name, 'meteorological equator', considers a number of important features of the zone of maximum development (Flohn, 1969). In this chapter, where the circulation is described, the term Inter Tropical Convergence Zone will be used, except when the zone is at latitudes over 23½°.

Thermal explanations of this zone point to the general similarity in position of the area of low pressure and the thermal equator (Figures 3.4, 4.2). Also, the low pressure is generally limited to the lower layers of the troposphere, rarely reaching levels over 700 mbar (3000 m), and this supports the idea of an origin near the earth's surface. On the other hand, latitudinal temperature differences over the equatorial oceans are extremely small, and it seems unlikely that they alone can generate a broad and persistent belt of low pressure. There are also some important differences in position between the thermal equator and the centre of low pressure, for instance over the central Pacific Ocean in January and over the western Pacific in July. These discrepancies point to other factors of origin.

Dynamic theories see the instability of the equatorial air masses as one of the main factors. The general convergence in the area creates a strong instability, which in turn reinforces the upward movement of air. This viewpoint is supported by the great width of the central zone. Convergence is not only caused by the confluence of air streams from the two Hadley cells, it also occurs within

Figure 4.3 The Inter Tropical Convergence Zone over eastern Africa shows a large number of towering cumulus clouds. (Between Dar es Salaam and Lusaka, February 1970, around noon local time)

these air streams, when they slow down or change in direction, as often happens at very low latitudes because of the reduction and change of sign of the Coriolis force.

On the basis of information available at present, it is impossible to evaluate the relative correctness of these two groups of explanations and it seems reasonable to assume a multiple origin of the central zone, with both thermal and dynamic forces responsible. It is certain that longitudinal differences are strong, with thermal origin more dominant over the continents and dynamic factors prevailing over the oceans (Barry and Chorley, 1972, p. 146).

It frequently happens that the central belt, in a position near the equator, shows a division into three zones. The two outer zones usually exhibit strongest convergence, while the centre displays only scattered areas of weak convergence, often related to westerly winds (Figure 4.4). These equatorial westerlies usually occur in rapidly moving zones and they are therefore often obscured in maps of

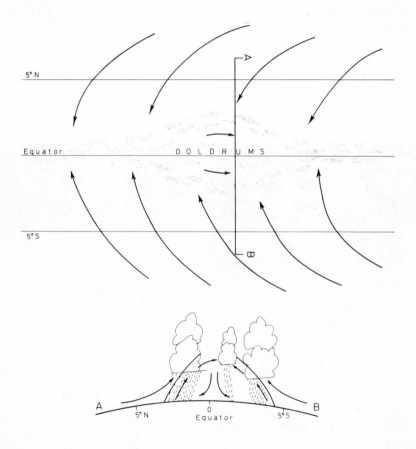

Figure 4.4 Generalized map and cross-section of the I.T.C.Z. in a position near the equator. Map: Shaded areas indicate precipitation. Cross-section: vertical exaggeration about 85 times. (After Fletcher, 1945)

the mean circulation pattern (Flohn, 1949). Most of the time, however, the winds are of variable direction and wind speeds are generally low. These are the 'doldrums' of the equatorial regions. When easterly winds prevail near the equator, subsiding air movements may occur, related to cooling of air, both by evaporation at lower levels and radiation losses to outer space near the top of the clouds (Asnani, 1967).

Over the whole Inter Tropical Convergence Zone weather conditions are generally characterized by the frequent occurrence of thunderstorms. The slightest surface heating or any other source of lifting of the very humid and warm air masses is sufficient to initiate vigorous convection. Larger disturbances occur occasionally, but they are usually of rather short duration, because they are rapidly filled up by compensating currents, which are not deflected by the Coriolis force near the equator.

Rainfall in the Inter Tropical Convergence Zone, when it is located near the equator, is characterized by very local distribution: large amounts fall over small areas, sharply delimited. This is related to the origin of the rainfall, mainly from convection cells. If the zone is split, most rainfall is usually received in the outer areas, where convergence is strongest. In the central parts most rainfall is associated with westerly winds (Flohn, 1960). The areas of maximum rainfall often do not coincide with the location of the main surface discontinuity, because it makes only a gentle slope with the horizontal (Figure 4.4). However, there are many variations from this generalized picture and every displacement of the convergence zone produces a different rainfall distribution pattern.

The identification of the convergence zone on weather maps is rarely possible from air mass characteristics, since converging air masses near the equator are very similar in temperature and humidity conditions. The convergence area is therefore usually identified by the thunderstorm activity around it. On satellite pictures the zone can be recognized easily by this same feature (Johnson, 1969). It has been calculated that a few thousand thunderstorms of the type prevalent near the convergence zone are sufficient to maintain the heat budget of the equatorial areas (Riehl and Malkus, 1958).

The mean position of the Inter Tropical Convergence Zone shows only minor seasonal variations over the oceans but large latitudinal movements over the continental areas (Figure 4.2). These displacements are clearly related to surface conditions, because the convergence zone follows the apparent movement of the overhead sun, and the accompanying belt of high temperatures, rather closely (Figure 3.4). With distance from the equator, certain changes in the character of the convergence zone takes place. The whole area of convergence becomes more diffuse and it attains a considerable width. Large variations in intensity occur in the various parts of the zone, but the general weather characteristics are about the same as near the equator. When the main convergence zone is well away from the equator, a secondary zone of convergence is often found close to the equator, where the wind directions of the trades are undergoing rapid changes because of the reversal in sign of the Coriolis force (Figure 4.2).

Over the oceans the mean position of the Inter Tropical Convergence Zone is generally a few degrees north of the equator, throughout the year. This is due to the more vigorous atmospheric circulation in the southern hemisphere, and the stronger cold ocean currents, as ocean basins are larger and more conducive to the formation of these currents than in the northern hemisphere.

In warm air masses, the decrease of pressure with elevation proceeds at a lower rate than in cool air. Therefore the relatively low pressure of the central zone of the tropical circulation slowly disappears in the higher parts of the troposphere. At the 500 mbar level, about 5600 m a.s.l., the pressure is generally higher than over the outer tropics. The upper troposphere shows a general outflow of air from the central zone of the tropical circulation.

The 'anti-trades'

In the classical theories of the tropical circulation cell, the term 'anti-trades' was used to indicate the upper tropospheric outflow of air. These currents are mainly found at elevations of 8000 m to 12 000 m, near the equator, slowly descending to about half that height at 25–30 degrees of latitude. They are free from surface friction and, moving generally from low to higher latitudes, they come strongly under the influence of the rapidly increasing Coriolis force. They are therefore deflected and become geostrophic westerlies.

The name 'anti-trades' is an unfortunate one, since it implies a similarity in character to the trade winds. As more and more upper air data became available, it was clear that a continuous and regular anti-trade wind system, comparable to the low level trades, does not exist in this form (Riehl, 1950, p. 12). Similarly, it was discovered that a pressure gradient between the equator and the areas of around 20 degrees latitude at the 250 mbar level, which would steer such a current, is often absent. Many observations showed no currents in polar and westerly directions at this level, but frequently indicated quite different winds.

The general picture of the tropical circulation at higher levels which is now emerging from many contradictory and confusing generalizations, shows large differences with longitude and strong seasonal variations. The best resemblance to the old simplified model of continuous and regular anti-trades is found over the eastern parts of the ocean basins, especially over the southern hemisphere and during the winter half-year. Over continental areas much more interrupted upper air movements prevail.

The subtropical jet stream consists of strong westerly winds at the 200 mbar level (around 12 000 m), which prevail throughout the year over 25–30 degree southern latitude. However, a similar jet stream over the northern hemisphere is limited to the winter season, and it is also found at a slightly higher latitude (Krishnamurti, 1972). During the summer it is replaced by a strong easterly current over Asia and Africa, at a latitude of about 10° N, the Tropical Easterly Jet (Flohn, 1964; Koteswaram, 1958). Since this current is obviously related to the monsoonal circulation at lower levels, it will be described more fully in the next chapter. Over the southern hemisphere easterly winds prevail during January at the 200 mbar level between 10 and 20 degrees latitude over Africa,

South America and northern Australia (Van Loon, *et al.*, 1971). These occurrences demonstrate the extent and size of the deviations from the old generalized picture of a continuous anti-trade circulation.

Where the anti-trades do prevail, they become almost purely westerly currents at a latitude of about 20 degrees, and there is no evidence of a further poleward movement of air in the upper parts of the troposphere at higher latitudes. A subsiding movement of air prevails generally over oceanic areas and over the continents during the winter half year.

The subtropical highs

Between the latitudes of about 20 to 40 degrees, the mean surface pressure maps are dominated by a number of elliptical areas, elongated in the east–west direction, with relatively high pressure, usually called the subtropical highs (Figure 4.2). They persist throughout the year over the large ocean basins, where they show only minor seasonal changes in position. They do fluctuate in intensity and size, however. In the southern hemisphere they display higher pressure during the winter seasons, and at that time they also extend their influence over the adjacent continental areas, so that an almost closed belt of high pressure is formed (Figure 4.2). During the summer, this belt is interrupted by lows over the continents. In the northern hemisphere the seasonal variations are different: over the oceans the subtropical highs display higher pressures and cover larger areas during the summer. In winter they are often connected with the continental highs at higher latitudes by high pressure ridges, but no continuous belt results.

The semi-permanent character of the subtropical highs, as illustrated by monthly mean pressure maps, is not always confirmed by daily weather charts, which show frequent variations in size, intensity and location of the high pressure cells.

As in all anticyclonic areas, the main air movements in the subtropical highs are downwards, and horizontal wind velocities are generally low. The subsiding movements create stable and dry air masses and rainfall is therefore extremely limited in areas which are frequently under the influence of the subtropical highs.

The subtropical highs often extend to rather high levels: they can still be identified occasionally at the 300 mbar level (about 9600 m a.s.l.)

An important feature of the subtropical highs is the asymmetry in their internal structure. Near the earth's surface the centre of highest pressure is generally over the eastern parts of the ocean basins, but at higher levels the core of maximum pressure is further to the west (Figure 4.5). The plane of the orbital circulation from the centre is therefore not strictly parallel to the earth's surface, but sloping gently upwards towards the west. Consequently, subsiding movements prevail mostly in the eastern half of the cell, but rising air currents occur frequently in the western sections. The result is that the western half of the cell has more unstable and humid air masses, yielding more rainfall than in the eastern half, where very stable air masses prevail almost continuously.

44

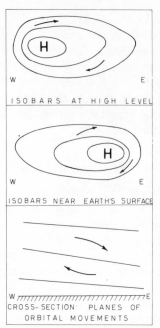

Figure 4.5　The vertical structure of a typical subtropical high pressure cell

The origin of the subtropical highs is rather complex and still a matter of considerable controvery. The classical theories explained the high pressure cells simply as the result of the piling up of poleward moving air in the anti-trades, when these had become completely deflected into westerlies, at a latitude of about 20 degrees. Convergence at high levels would result in downward movement of air and high pressure near the earth's surface. The weak point in this dynamic explanation is, of course, the frequent absence of the anti-trades, which makes it impossible to attribute the origin of the rather persistent subtropical highs to this one factor. Another dynamic theory assumes that polar air masses are the main cause of the subtropical highs. Because of the changes of the Coriolis force with latitude, anticyclonic cells near the Polar Front show a tendency to move equatorwards, while cyclones generally move polewards (Rossby, 1947). These moving cold anticyclones therefore regenerate the subtropical highs frequently. This explains the pulses in the intensity of the subtropical highs observed on series of daily weather maps. The polar outbreaks would generally show a preference for the eastern parts of the ocean basins, where cold ocean currents prevail, and where friction along the continental coasts, especially in the southern hemisphere, give a strong meridional impetus to these movements. This preference explains the higher pressure over the eastern oceans at low levels and the generally stronger development of these centres over the southern hemisphere (Pédelaborde, 1958, pp. 35–45).

Thermal theories also consider two different areas of origin. Upper air cooling takes place as the air masses of the anti-trades lose heat to outer space by radiation. These air masses therefore become denser with growing distance from the equator, resulting in higher pressure at all levels. A second area of cooling is found near the earth's surface: the cold ocean currents and the cool continents during the winter. This second source explains the cellular pattern of the subtropical highs and their extension over the continents in winter very well.

It seems probable that the subtropical highs are of multiple origin, with both dynamic and thermal factors, near the earth's surface as well as at higher elevations, responsible for their formation. Sometimes these factors reinforce one another, at other seasons or locations they have the opposite effect and tend to cancel out (Barry and Chorley, 1972, p. 155). These variations explain the changes in intensity and location of the subtropical highs.

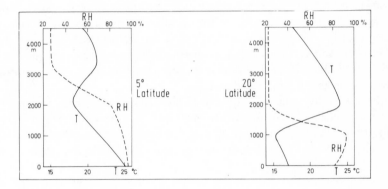

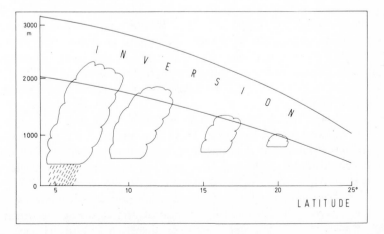

Figure 4.6 A generalized cross-section in meridional direction of a trade wind. Top: vertical temperature and humidity profiles at 5 degrees and 20 degrees latitude.

The trade winds

Between the subtropical highs and the Inter Tropical Convergence Zone the low level general circulation over the oceans in both hemispheres is dominated by strong and persistent easterly winds, the trades. These winds cover very large areas, of about 20 degrees of latitude in the summer hemisphere and approximately 30 degrees in the winter hemisphere (Figure 4.2). At any time, therefore, more than a third of the earth's surface is under the influence of trade winds (Crowe, 1949, 1950).

Generally the trades are best developed over the eastern parts of the ocean basins, while over the continents and the western oceans they are frequently interrupted by winds of different directions (Pédelaborde, 1958, p. 38).

The trades are mighty air currents, in which usually three distinct layers can be identified (Figure 4.6). The *lower trades* are limited to elevations of about 500 m near the subtropical highs, but they increase in thickness on the way towards the equator and near the Inter Tropical Convergence Zone they can reach levels as high as about 2500 m. In this layer the winds are strong and very steady, qualities which gave the trades their name during the days of the sailing vessels. The general direction of the lower trades is ENE in the northern and ESE in the southern hemisphere and generally over 80 per cent of all observations indicate winds from within 45° of this main direction. Over the oceans these constancy figures can be as high as 95 per cent.

In the lower trades the wind speeds are also very regular. They vary between means of about 10 km/h during the summer and approximately twice that speed during the winter. Only near the subtropical highs does the speed of the lower trades change more frequently, in accordance with pulses in the intensity of the high pressure cell.

The air masses of the lower trades are usually humid and their temperature is governed by the sea surface water temperatures of the ocean, and therefore generally increase towards the equator. Because of the shallowness of the lower trades, the prevailing cloud type is the short cumulus type, cut off at the top of the layer and leaning slightly backwards, because wind speeds are highest near the surface (Figure 4.6). These 'trade wind cumuli' rarely produce any precipitation, except near the equator, where they grow much taller, and climates under the domination of the trades are therefore characterized by rather dry conditions.

Over the lower trades lies the trade wind *inversion layer*, in which the temperature increases with elevation. The rate of increase is about 1 °C/100 m at 20° latitude (Figure 4.6). The thickness of the inversion layer varies from a few hundred to about a thousand metres, and increases as the trade wind proceeds towards lower latitudes. At the same time, the temperature difference between the bottom and the top of the inversion decreases from about ten to only a few degrees. The inversion is generally strongest during the winter and best developed over the eastern parts of the ocean basins (Riehl, 1954, pp. 53–70).

The main effect of the trade wind inversion is that it acts as an efficient lid, which prevents all upward movements of air. This is clearly illustrated by the cloud formations of the lower layer (Figure 4.6). As a result, precipitation in the trade wind areas is rare, except where the inversion is lifted or destroyed. This can be achieved by disturbances of the 'easterly wave' type (Chapter 5), or by orographic lifting.

The inversion layer disappears when convergence prevails. This happens where the trades change direction or slow down in speed, as is the case near the geographical equator and, of course, near the Inter Tropical Convergence Zone.

The *upper trades*, above the inversion, are dominated by geostrophic easterlies, which generally have much lower wind speeds than the lower trades. The upper trades reach elevations of about 6000 m near their origin, rising to about 10 000 m near the Inter Tropical Convergence Zone. As heat and moisture originating from the earth's surface rarely penetrate through the inversion layer, the air masses of the upper trades are generally dry and stable. Their direction is almost purely easterly.

Explanations of the trade winds lay emphasis either on temperature–pressure conditions, or on dynamic factors as the main origin. Thermal theories see the pressure gradient between the subtropical high and the equatorial trough as the major factor of the trades. Many observations support this view: the higher wind speeds near the earth's surface, the consistency of the lower trades, the variations of the trades with changes in the intensity of the subtropical highs are the most important. Quantitatively the pressure gradient is sufficient to maintain such a mighty air current. In this view, the inversion is explained by adiabatic warming of the air during the subsidence in the subtropical highs, followed by surface cooling of the lowest layers by the cool ocean waters or, in winter, the relatively cool continents. This is also supported by observations, especially that the inversion is best developed over the eastern parts of the ocean basins, where cool currents prevail.

Dynamic theories see cold polar air masses as the main source of the inversion (Pédelaborde, 1958, pp. 45–46). These cold air masses are propelled towards lower latitudes by centrifugal forces, because the radius of the earth's rotation is largest near the equator. The interruptions of the trades by easterly waves are, in this view, related to disturbances in the polar front region, creating the intermittent character of the invasions of polar air masses.

Again, it seems likely that both thermal and dynamic forces are responsible for the trade winds, though most observations seem to support the classical, thermal theories.

The main features of the general circulation over the tropics have now been described. This generalized pattern prevails, throughout most of the year, over the oceans, but over the continents important seasonal deviations occur. These are the subject of the next chapter.

48

References

Alaka, M. A., 1964, *Problems of tropical meteorology*, WMO Technical Note No. 62, Geneva, 36 pp.

Asnani, G. C., 1967, Subsidence near the Equator, *Nature*, **214**, 73–74.

Barry, R. G. and Chorley, R. J., 1972, *Atmosphere, Weather and Climate*, 2nd ed., London, Methuen, 379 pp.

Beckinsale, R. P., 1957, The nature of tropical rainfall, *Tropical Agriculture, Trinidad*, **34**, 77.

Crowe, P. R., 1949, The trade-wind circulation of the world, *Transactions and Papers, Institute of British Geographers*, **15**, 39–56.

Crowe, P. R., 1950, The seasonal variation in the strength of the trades, *Transactions and Papers, Institute of British Geographers*, **16**, 25–47.

Fletcher, R. D., 1945, The general circulation of the tropical and equatorial atmosphere, *Journal of Meteorology*, **2**, 167–174.

Flohn, H., 1949, Eine äquatoriale Westwindzone als Glied der allgemeinen Zirkulation, *Zeitschrift für Meteorologie*, **3**, 240.

Flohn, H., 1060, *The structure of the Intertropical Convergence Zone*, in: Bargman, D. J. (Ed.), *Tropical Meteorology in Africa*, Nairobi, Munitalp, pp. 244–246.

Flohn, H., 1964, *Investigations on the Tropical Easterly Jet, Bonner Met. Abh.*, **4**, 8.

Flohn, H., 1969, *Investigations on the atmospheric circulation above Africa, Bonner Met. Abh.*, **10**, 38–39.

Hadley, G., 1735, Concerning the cause of the general trade winds, *Phil. Trans. Royal Society*, London, **39**, 58.

Johnson, D. H., 1969, *The role of the tropics in the global circulation* in: Corby, G. A. (Ed.), *The global circulation of the atmosphere*, London, p. 113–118.

Koteswaram, P., 1958, The Easterly Jet Stream in the Tropics, *Tellus*, Vol. 10, p. 43–47.

Krishnamurti, T. B., 1972, *Tropical observations and zonal-time averages*, in Young, J. A. (Ed.), *Dynamics of the Tropical Atmosphere*, N.C.A.R., Boulder, Colorado, pp. 11–19.

Palmén, E., 1951, The role of atmospheric disturbances in the general circulation, *Quarterly Journal Royal Met. Society*, **77**, 337–354.

Pédelaborde, P., 1958, *Les moussons*, Paris, Armand Colin, 208 pp.

Riehl, H., 1950, On the role of the tropics in the general circulation of the atmosphere, *Tellus*, **2**, 1–7.

Riehl, H., 1954, *Tropical Meteorology*, New York, McGraw-Hill, 392 pp.

Riehl, H. and Malkus, J. S., 1958, On the heat balance in the equatorial trough zone, *Geophysica*, **6**, 534.

Rossby, C. G., 1947, On the general circulation of the atmosphere in middle latitudes, *Bull. American Met. Soc.*, **28**, 255–280.

Van Loon, H., Taljaard, J. J., Jenne, R. L., Crutcher, H. L., 1971, *Climate of the upper air: Southern hemisphere*, Vol. II — *Zonal Geostrophic Winds*, Figures 6, 10.

CHAPTER 5

Regular Variations of the Tropical Circulation

Many important variations of the generalized pattern of the tropical atmospheric circulation, which was described in the last chapter, occur regularly in seasonal or diurnal cycles. Climatologically most important are deviations which show a seasonal pattern, but those recurring with a diurnal frequency can have a strong effect on climates where they occur often.

Seasonal variations — the Monsoons

Over the large ocean basins, seasonal changes in the tropical circulation are limited to minor latitudinal shifts and small variations in intensity of the main components, but the general pattern remains virtually the same throughout the year. However, the picture is entirely different over the tropical continents and adjacent seas, where important changes takes place regularly with a clear seasonal rhythm.

The determinant effect of the continents in this respect is that they create much larger seasonal temperature variations than the oceans (p. 21). This is particularly the case at latitudes of about 15 to 30 degrees, which receive very large amounts of solar radiation during the summer months (Figure 2.4). Over the oceans the surplus energy is rapidly absorbed by the water surface and spread over great depths by waves and turbulence, and dissipated to other latitudes by ocean currents, so that no strong increase of surface temperatures result. But over the continents the incoming radiation is largely used to warm the earth's surface, so that temperatures in these regions reach very high levels, the highest regularly experienced anywhere on earth (Figure 3.4).

The main consequence for the general circulation is that the subtropical high pressure cells, which are permanently present at these latitudes over the oceans and extend over the continents during the winter, are rapidly transformed at the beginning of the summer into centres of low pressure over the tropical continents. These thermal lows then gradually take over some of the functions of the 'equatorial' trough, which changes in character near the equator, becoming less intense, and a new areas of intertropical convergence is established some distance from the equator (Figure 4.2). The trade winds from the winter hemisphere move across the equator, where they usually slow down and create a minor area of convergence, owing to their change in direction, caused by the reversal of the Coriolis force. Once in the summer hemisphere, they become quasi-geostrophic westerlies (Flohn, 1955).

During the winter, the continents in the indicated latitudes experience rather low temperatures, resulting in relatively high surface pressure and the re-establishment of winds towards the equator, similar to the trade winds over the ocean, though generally less constant in both speed and direction.

The tropical continents and adjacent oceans therefore experience a semi-annual reversal in wind direction, and the winds in these systems are usually called 'monsoons'. There are many different definitions of this term, and some of them include wind systems where the seasonal reversal of winds is largely statistical rather than actual, and caused by a difference in location of the polar front (Khromov, 1957; Pédelaborde, 1958, pp. 9–29). Generally, however, the word 'monsoon' is used only for wind systems where the seasonal reversal is pronounced and exceeds a minimum number of degrees. Here, we shall limit it to systems which show a seasonal change of wind direction of at least 120 degrees, while both winds must have a constancy higher than 40 per cent and a mean resultant speed of more than 3 metres/second (Ramage, 1971). The areas in which monsoons, according to this definition, prevail are very large and mainly situated in the tropics (Figure 5.1).

The above explanations of the monsoons represents the classical view. It emphasizes thermal conditions as their main origin and thereby explains the major common features of monsoon systems satisfactorily. However, many regional variations in this general concept are caused both by factors at the earth's surface, such as the form of the continents and their relief, and by conditions in the upper troposphere. It is therefore necessary to describe the various monsoons regionally.

The importance of the monsoons is enormous. This is illustrated by the fact that the area where climates are largely controlled by monsoons has a population of more than 2000 million, or about 54 per cent of the total world population. Moreover, most people in this region derive their income entirely from agriculture. Their lives depend on the monsoon rains, which are essential for food production as well as most cash crops. Any failure or even late arrival of these monsoon rains causes widespread starvation and economic disasters. It is mainly for this reason that the Asian summer monsoon especially has been studied widely.

The Asian monsoons

The seasonal variations in the general circulation are most pronounced over southern and eastern Asia, in a large area which stretches from Pakistan in the west to Japan and northern Australia in the east (Figure 5.1). The main reasons for this very strong development of monsoons are the sizes of both the Asian continent and the adjacent oceans, the largest in the world. Another favourable factor is the very high and extensive mountain backbone of the continent, which

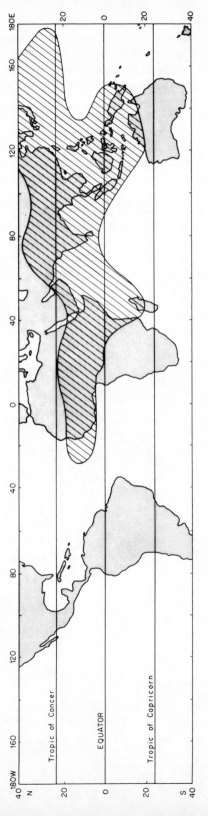

Figure 5.1   Areas with monsoon circulations (after Ramage, 1971, p. 4)

is elongated in a west–east direction at a critical latitude, thereby forming a barrier between tropical and polar air masses. The following explanation of the Asian monsoons is a condensation of a large number of studies, based on many different observations and theories, some of which are contradictory and rather controversial.

During the *winter* season of the northern hemisphere a high pressure centre of great intensity develops over the northern parts of the Asian continent, at latitudes of about 40–60 degrees, where very low temperatures prevail, mainly as the result of radiation losses from a largely snow-covered surface (Figure 3.4). The outflowing air masses spread towards the south and south-east over Korea, China and Japan. They converge, at a latitude of 15°–20°N, over the South China Sea with the north-easterlies from the Pacific Ocean. The boundary between these two air streams is frequently considered as a branch of the Polar Front, since the continental air masses are still cold in comparison to the maritime air from the Pacific. The two air currents merge gradually on their way to the southwest, where they form the north-east monsoon of Malaya. They recurve into westerlies upon crossing the equator, over Indonesia. The Inter Tropical Convergence Zone is at a latitude of about 5°–10° S, but a branch of the monsoon current continues towards a strong surface heat low over northern Australia (Figure 5.2).

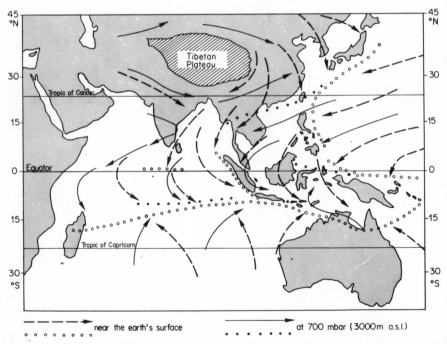

Figure 5.2   The Asian winter monsoon. Winds (arrows) and convergence zones (dots) from about December to March, near the earth's surface and at the 700 mbar level (about 3000 m)

The circulation during the same season at 700 mbar (about 3000 m a.s.l.) shows that the Asian winter monsoon is rather shallow near its centre of origin, which fact confirms its predominantly thermal causes (Figure 5.2.). Over most of the Asian continent, winds at the 700 mbar level are westerly, and their main path is to the south of the Tibetan–Himalayan massif. On the eastern side of the Highland plateau the two branches of the westerlies reunite in a zone of confluence, which reaches as far as southern Japan. This zone often gives rise to depressions and it is regarded as an upper-air branch of the Polar Front. Further south, where the winter monsoon is reinforced by the north Pacific trades, it reaches to the 700 mbar level. Over northern Australia the monsoon circulation is again rather shallow, as easterlies prevail at 3000 m.

While most of southern and eastern Asia is strongly under the influence of the winter monsoon, northern India is an exception. The winter monsoon is rarely experienced over this area, where the winds are mainly from westerly directions. The main reason is the mountain barrier of the Himalayan ranges, which prevents the polar air from moving over the Ganges lowlands. Southern India, south of about 20° N, experiences the winter monsoon, but it is rather weak and its air masses are not of polar origin. In India, therefore, the term 'monsoon' is normally used only for the summer winds, and more specifically for their accompanying rainfall.

The air masses of the Asian winter monsoon are of two different origins. In northeastern Asia they are polar–continental: dry, cold and stable; and northern China and Korea consequently experience rather cold and dry winters. On their way towards Japan and southern China these air masses are rapidly modified, at least in their lower layers, by contact with the relatively warm sea surfaces. They bring less cold winters, with snow in the northern areas, over Japan and central China. A further modification takes place over the South China Sea, where the originally continental air masses are mixed with the second type: warm and stable air masses from the Pacific Ocean. As a result of the many depressions formed in this area, the marine air masses are uplifted, and only traces are left of the original trade wind inversion when they reach Indo-China and Malaya, where it is usually impossible to identify the air masses by area of origin (John, 1950).

The combined monsoon air masses become truly equatorial over Indonesia: they are then very humid, warm and unstable. They keep these characteristics as far as northern Australia.

Summer conditions over Asia are much more complicated, but because of the great importance of the summer monsoon as the main source of precipitation in many parts of the area, more studies and observations are available than for the winter monsoon (Jen-Hu Chang, 1967; Lockwood, 1965).

The main thermal origin of the summer monsoon is beyond doubt, but its development at the beginning of the summer season shows that the idea of a single thermal low over the continent is an oversimplification. This low is situated over northwestern India, but the monsoon develops first over southern China, then progresses to Burma and does not start over India until more than a

54

month later (Barry and Chorley, p. 277). The main cause of this retardation of the monsoon over India is the upper air circulation, at a level around 6000–8000 m. At this elevation, westerly winds prevail over the whole area south of the Tibetan–Himalayan massif during the winter, conditions being very similar to those at about 3000 m (Figure 5.2). This circulation causes a meridional upper air trough, situated approximately over the Bay of Bengal. With the beginning of the summer, this trough favours the development of upper air easterlies over Indo-China and southern China. These easterlies seem to be a necessary condition for the establishment of the summer monsoon (Yin, 1949). They constitute the upper return movement of air towards the equator and are therefore part of the monsoonal circulation during the summer over Asia (Figure 5.3). Over northern India, however, the prevailing westerlies at the 8000 m level prevent the development of the summer monsoon. Usually about the end of May, the upper air westerlies suddenly shift to a path north of the Tibetan plateau, the meridional upper air trough moves to a more westerly position, at about 75° E, and upper-air easterlies can establish themselves over northern India, opening the way for the summer monsoon at lower levels (Koteswaram, 1958).

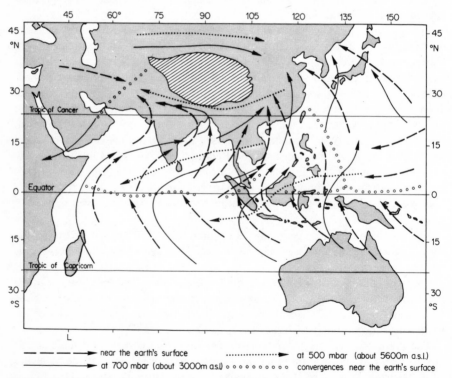

Figure 5.3   The Asian summer monsoon. Winds (arrows) and convergence zones from about June to September, near the earth's surface. Prevailing winds at 700 mbar (300 m) and at 500 mbar (about 5600 m)

55

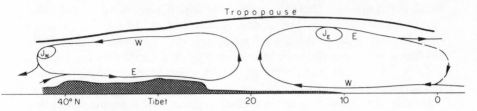

Figure 5.4 A meridional cross-section of the Indian summer monsoon (after Koteswaram, 1958).
J_E — Tropical Easterly Jet; J_W — Westerly Jet Stream. Vertical exaggeration: about 500 times

The upper-air easterlies associated with the Asian summer monsoon culminate in the Tropical Easterly Jet. This current is usually situated at an elevation of 12 000–15 000 m, its core is at a latitude of about 15° N and it can attain wind speeds of 120–150 knots (Flohn, 1964). The Tropical Easterly Jet prevails from about June to September over Asia and Africa. During this period it replaces the southern branch of the Westerly Jet Stream, which dominates the circulation at this level during the northern hemisphere winter (Koteswaram, 1958).

The large mountains and highlands of Central Asia favour the development of the Indian summer monsoon, not only by their deflection of the upper-air westerlies to more northerly paths, but also as they are an important source of heat in the higher levels of the troposphere. The very intense solar radiation in the highlands creates high temperatures and reinforces the low pressure centre over northern India (Flohn, 1960). A meridional cross-section emphasizes the thermal structure of the Indian summer monsoon and its relation with the Tibetan–Himalayan massif (Figure 5.4).

Because of these rather favourable conditions, the Asian summer monsoon is a mighty air current. Over India it reaches up to about 6000 m while over Burma it can be as high as 9000 m during June and July (Figure 5.3). Further to the east, where the influence of the mountains and highlands is much less, the monsoon current is shallower; over Japan it rarely reaches levels over 2000 m (Pédelaborde, 1958, p. 107).

Constancy figures of the Asian summer monsoon regarding both speed and direction are generally lower than for comparable trade winds. The monsoon frequently exhibits a pulsating character, with changes at intervals of about 3–10 days between a very active monsoon current and an almost complete break in all monsoon activities (Jen-Hu Chang, 1967, pp. 386–387).

The main importance of the Asian summer monsoon is related to the precipitation which it brings. In many parts of southern and eastern Asia most of the annual rainfall is received during the summer (Das, 1972, p. 16; Barton, 1962; Kendrew, 1937). The monsoon rainfall is closely related to the characteristics of its air masses,. These are from three main regions. West of about 100 degrees eastern longitude the main source region is the Indian Ocean south of the equator. These air masses have moved over the warm ocean surface for thousands of miles and they are consequently very humid, warm and unstable. All traces of the original trade wind inversion have disappeared near the equator, where these air masses crossed a secondary convergence area (Figure 5.3). East of 100° E the main source region of the monsoon air masses is the subtropical high pressure cell over Australia. These air masses are originally stable and dry and they keep these characteristics over the southeastern islands of Indonesia, but further westwards they become more humid and unstable.

The main boundary between these two air masses is frequently over the Malay peninsula, where it causes much rainfall. Its exact location is often rather difficult to establish, because the two air masses have very similar temperature and humidity conditions.

A third air mass of the Asian summer monsoon comes from the Pacific Ocean. This air mass, humid and warm but originally rather stable, forms the branch of the monsoon which covers Japan and northern China. On its way over the warm ocean surface it is rapidly modified and becomes unstable (Lautensach, 1950). The zone of confluence with the second air mass is generally to the east of the Philippines, but can reach southern Japan and even Korea (Figure 5.3).

It is obvious that all these air masses, humid and unstable after their long journey over warm ocean surfaces, can yield large amounts of precipitation. The actual processes involved are orographic lifting, convergence and disturbances (Das, 1972, pp. 113–115, 124–130). The monsoon disturbances generally travel with the monsoon, They are clearly related to the pulsations in the strength of the main current, and over flat areas they are probably the major source of monsoon rainfall (Barry and Chorley, 1971, p. 281). Usually they bring a few days of heavy rainfall, followed by interruptions, lasting 7–10 days, with a rather sluggish monsoon current.

As the monsoon rainfall is determined by the interplay of various factors, its prediction is still rather difficult (Das, 1972, pp. 139–144). Rainfall variability is also high for this same reason, a factor of eminent importance for the people of southern and eastern Asia.

In the very large area of the Asian monsoons considerable regional and local differences exist. Generally the winter monsoon is stronger in the eastern parts of the region, while the summer monsoon is better developed in the west. In south-east Asia, near the equator, the monsoons are of about equal intensity.

Australian monsoons
The monsoonal circulation over northern Australia constitutes an extension of the Asian system, but with seasonal characteristics reversed. The Asian winter monsoon becomes the north-west monsoon of northern Australia, which has the summer season during its prevalence and, as it brings rather warm, humid and unstable air masses to the area, it has the true characteristics of a summer monsoon (Figure 5.2). During the southern hemisphere winter the main wind direction is southeasterly and the origin of these winds is the high pressure cell

over the southern Pacific, which has extended over a large part of the Australian continent (Figure 5.3). These winds are basically trade winds and they are bringing dry and stable air masses to northern Australia. Though there are no polar air masses involved, these winds still have the main features of a winter monsoon.

African monsoons

The monsoonal circulations over Africa differ from the Asian system in two important ways. First, their magnitude is much smaller, not only regarding the areas covered, but also in respect of the thickness of the air layers involved (Figures 5.5 and 5.6). The main reason for this smallness is the limited seasonal variation in the latitudinal position of the main elements of the general circulation, which amounts to not more than 15 degrees of latitude over western Africa and about 30 degrees over East Africa, compared to about 40 degrees over monsoon Asia. Secondly, no polar air masses are involved in the African monsoons and the monsoons differ much less in the main characteristics than in Asia. It is for this reason that some authors do not consider the African system as truly monsoonal (Pédelaborde, 1958, pp. 172–174). It is certain, however, that a genuine seasonal reversal of wind directions exists over large parts of tropical Africa, in which both winds are well above the minimum limits of speed and constancy indicated earlier (p. 50).

The African monsoons are surface winds, which rarely reach levels over 5000 m (Figures 5.5, 5.6). Their most important feature is a regional differentiation: in West Africa there is a good deal of difference between the characteristics of the two monsoon winds, but in East Africa the two winds differ only in direction and the air masses which they bring are remarkably similar. The main reason for these regional variations is, of course, the form of the African continent. In West Africa a large continental area north of the equator contrasts with the marine regions of the South Atlantic Ocean, but in East Africa the continent stretches on both sides of the equator, though it is broader in the northern hemisphere. Other factors which accentuate the differences between West and East Africa are the large meridional mountain ranges related to the Great Rift Valley and the influences of the Asian monsoons, experienced only in East Africa.

As large parts of Africa consist of extensive highlands, the circulation near the earth's surface is represented by the 850 mbar level, corresponding to about 1500 m above sea level. At this elevation, most air streams are relatively free from direct surface effects, such as friction (Figures 5.5; 5.6).

In *West Africa* the continent is largely under the influence of northeasterly trade winds during the northern hemisphere winter, with the Inter Tropical Convergence Zone close to the equator (Figure 5.5). The northeasterlies prevail to an elevation of about 3000 m and bring dry and stable air masses, which often carry dust particles from the desert regions over which they originate. These winds are locally called the 'harmattan' in many parts of West Africa. Except for a narrow area along the southern coast, this is the dry season in West Africa. During the northern hemisphere summer, high temperatures prevail over the

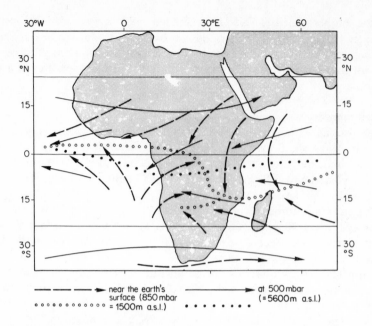

Figure 5.5 The circulation over Africa during January at the 850 mbar and 500 mbar levels (arrows: winds; dots: convergence zones)

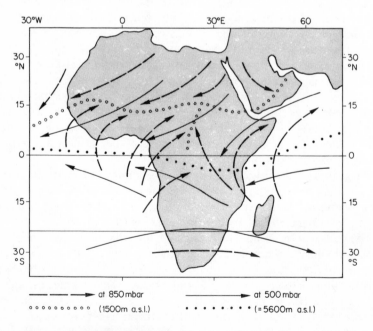

Figure 5.6 The circulation over Africa during July at the 850 mbar and 500 mbar levels (arrows: winds; dots: convergence zones)

continent and a thermal low pressure area builds up at about 20 degrees northern latitude (Figure 4.2). Consequently the Inter Tropical Convergence Zone, often locally termed the Inter Tropical Discontinuity, slowly moves northwards to reach a position which at the earth's surface is around 15° N (Figure 5.6). Southwesterly winds invade the continent and these are generally referred to as the 'monsoon'. The air masses which the monsoon carries are warm and humid. Most rainfall is usually received not near the surface position of the air mass discontinuity, but further south, because the continental air masses are warmer than the oceanic ones, so that they are uplifted. As only the oceanic air masses produce the precipitation, the zone of maximum rainfall is where these air masses are thicker (Figure 5.7). The West African monsoon is relatively shallow and at the 500 mbar level the Inter Tropical Discontinuity remains near the equator throughout the year (Figures 5.5; 5.6) (Leroux, 1973).

In *East Africa* continental influences are effective on both sides of the equator and consequently the latitudinal shift of the Inter Tropical Convergence Zone is much larger. In January it is situated at about 15° S and most of East Africa is under the influence of northeasterly winds, which become northwesterlies south of the equator (Figure 5.5). These winds are largely of continental origin and they produce little rainfall over East Africa, also because widespread divergence prevails. The only areas with precipitation during this season are situated near the main zones of convergence.

In July, the Inter Tropical Convergence Zone is at about 15° N, and over East Africa southeasterly winds prevail. Again, these winds bring little precipitation. The air masses which they bring to East Africa are partly of continental origin, coming from the high pressure area over South Africa. But even those air masses which come from the Indian Ocean are dry, as they have shed most of their moisture on the steep eastern mountain slopes of Madagascar. Therefore, in most of East Africa both monsoons are relatively dry, and rainfall is concentrated during the intermediate seasons, when the Inter Tropical Convergence Zone moves over the region on its way to the opposite hemisphere. As in West Africa, there is no trace of any monsoonal circulation at the 500 mbar level (Figures 5.5, 5.6).

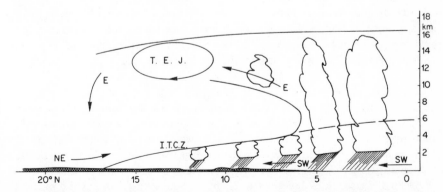

Figure 5.7 Meridional cross-section of the troposphere over West Africa during August (after Leroux, 1973). Vertical exaggeration: about 50 times

South American monsoons

Though there is some seasonal reversal of wind directions over South America, it is not included in the monsoon regions (Figures 4.2, 5.1). This is because the prevailing winds do not meet the standards of average speed and constancy used here. The main reason is, of course, the relatively small size of the continent and its limited extent in the northern hemisphere. Therefore strong and persistent thermal low pressure centres do not develop over the continent. Moreover, the western coastal areas are bordered by a rather cold ocean current, even at equatorial latitudes, so that the continent remains warmer than the adjacent ocean to the west throughout the year, preventing a high pressure cell from developing over the continent during the winter season of the southern hemisphere.

Diurnal variations of the general circulation

Diurnal wind systems are only of climatological importance where they occur frequently and regularly. This is the case in many tropical areas, where diurnal temperature variations are generally more conspicuous than in the mid-latitudes (page 21). The thermal changes between day and night, so typical of the tropics, are the main driving force of the diurnal wind systems, because they differ in intensity over land and water surfaces, and over highlands and lowlands. Other favourable circumstances are the generally small pressure gradients and low wind velocities of the general circulation, which reduce large-scale turbulence and allow the rapid formation of local pressure differences. The absence of fronts and strong depressions makes the development of diurnal wind systems a rather regular phenomenon in the tropics, and they are an important feature of many tropical climates.

Because of their limited duration, diurnal wind systems usually are effective only over relatively small areas, where the main causes of their development prevail. They rarely extend far from their regions of origin and by their very nature show many local variations.

Generalizing broadly, there are two main types of location of diurnal winds: coastal regions, both along the sea and near large lakes, where systems of land and sea (or lake) breezes occur frequently, and areas of variable relief, where different types of valley and mountain winds can develop.

Sea and land breezes

Coastal wind systems which a clear diurnal cycle are not limited to the tropical climates, but they generally show their most regular occurrence and strongest development here. Thermal differences between land and water surfaces are their main cause. During the day, the land heats up rather quickly under the influence of solar radiation, while water surfaces remain cooler, because the heat is dissipated over thick layers of water by turbulence and waves, and by direct penetration and absorption. As a result, a small convectional system develops,

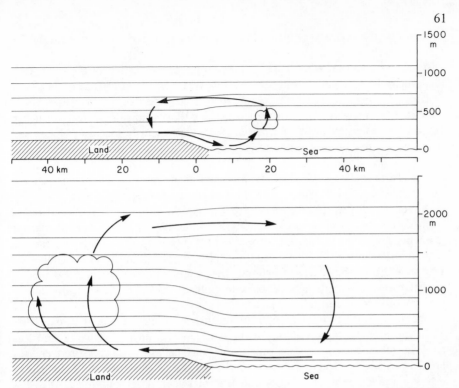

Figure 5.8 The basic pattern of sea and land breezes. Top: At night: land breeze. Bottom: During the day: sea breeze.
Thin lines represent isobaric surfaces. Vertical exaggeration: about 20 times

with winds near the earth's surface blowing towards the land — the sea breeze (Figure 5.8). At night, the land cools off rapidly, while the water surface remains at about the same temperature as during the day, and a reverse pressure difference results in the land breeze (Riehl, 1954).

The *sea breeze* is usually the stronger of the two winds. It can, under favourable conditions, reach speeds of 8–15 knots, and the thickness of the air layer involved can be as much as 1000 m. The sea breeze can reach inland as far as 50 km, but it cannot always by identified clearly over these distances, as it may be mixed up with other local winds. The return flow of the sea breeze is usually at levels between 1500 m and 3000 m.

The sea breeze usually starts, near the beach, a few hours after sunrise and it then progresses inland. It attains its maximum development during the early afternoon. It normally continues until shortly after sunset, but its circulation at higher levels may persist for a few hours longer (Kimble, 1946, p. 101).

Seasonally, the sea breeze is strongest when insolation is intense. It is therefore best developed during dry seasons. In the outer tropics the summer is also a season of strong sea breezes, if cloudiness is not heavy. The doldrum areas are usually favoured with strong breezes because the low wind velocities of the general circulation and the unstable air masses create propitious conditions for · their development.

As with all diurnal winds, the actual strength and direction of the sea breeze are controlled by local factors. Low surface water temperatures, caused by cold ocean currents or upwelling of water from below, reinforce it. Factors which increase the daytime temperatures over land have the same effect. A dense vegetation cover, swamps or flooded ricefields generally lead to reduced sea breezes. The presence of mountains near a coast often create a combined sea-breeze–valley-wind system (Riehl, 1954).

The sea breeze rarely brings much precipitation, because its air masses have experienced a stabilizing downward movement over sea to begin with. But when it converges with winds from a different direction, often a sea breeze 'front' is formed and this can cause local rainfall (Ramage, 1964, p. 66). Over islands and peninsulas, systems of converging sea breezes from opposite coasts can cause a regular afternoon rainfall maximum (Byers and Rodebush, 1948; Nieuwolt, 1968). During the early morning, the beginning sea breeze can carry disturbances, which developed over sea during the night, to coastal areas, thereby producing rainfall.

The *land breeze* is weaker than the sea breeze in most tropical climates. Its main cause is the rapid cooling of the land surface during the night, and this cooling influence is limited to only a thin surface layer of air. Moreover, this layer is strongly affected by friction (Figure 5.8). Therefore the land breeze rarely exceeds speed of 5 knots and the thickness of the moving air layer is usually only a few hundred metres. The land breeze does not normally reach more than about 15 km seaward (Kimble, 1946, pp. 100, 113). The land breeze generally starts about three hours after sunset and then increases in strength until sunrise, and it can continue for a few hours after that.

Land breezes are best developed in areas where the water surface is relatively warm: in equatorial regions, near warm ocean currents and at relatively low shallow lakes, such as Lake Victoria. Long and clear nights, occurring during dry seasons and, in the outer tropics, during the winter, are also favourable for land breezes.

All breezes are strongly influenced by the winds of the general circulation. Where these are strong, no breezes develop at all, since turbulence prevents the establishment of local temperature and pressure differences between water and land surfaces. With weaker general winds, the breezes are often limited to changes in the direction and speed of these winds (Nieuwolt, 1973, pp. 190–195). In doldrum areas and near the equator it can happen that the breezes dominate locally over the weak general circulation (Wexler, 1946).

Other variations of the breezes are related to the general form of the coastline, which can cause local convergence or divergence (Watts, 1955). Breeze systems develop over islands, when these are not too small. A diameter of about 15 km seems to be the minimum. Over narrow seas, such as the Straits of Malacca, systems of converging land breezes may develop at night, creating precipitation (Ramage, 1964, p. 64).

As with most local winds, breezes are generally not affected by the Coriolis force, unless they prevail over large distances. In the outer tropics this factor might cause a deviation of the breezes, which can become parallel to the coastline, but this is rarely the case at low latitudes.

Sea and land breezes are of great practical importance. In many tropical lowland areas they bring very welcome relief from the oppressive heat during the hottest hours of the day (Nieuwolt, 1973, pp. 199–205). This is caused both by the advection of cooler air and by improved ventilation. Fishermen also use the breezes: they move out to sea with the land breeze in the early morning and return to land with the sea breeze in the afternoon.

Mountain and valley winds

Over areas with large differences in relief, diurnal wind systems develop frequently, and these winds are particularly regular and strong in the tropics. Their basic origins are temperature differences between points at the same elevation, but with a different distance from the earth's surface. During a sunny day the highlands heat up rapidly under the influence of the intense insolation, but the free atmosphere, over the lowlands, is much less affected. An upward air movement, directed towards the highlands, is the result (Figure 5.9). This is the valley or anabatic wind, which can easily be recognized as it is often accompanied by the formation of cumulus clouds near the mountains or over escarpments and slopes. At night, a reverse temperature difference develops, as highlands cool off rapidly because of terrestrial radiation losses. A form of air drainage towards the lowest points take place. This is the mountain wind, or generally the katabatic wind (Figure 5.9).

These wind systems can develop in largely different dimensions: over a single mountain or valley, or even an individual slope; along mountain ranges or

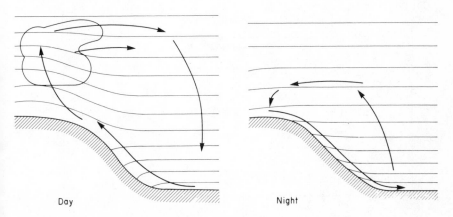

Figure 5.9   The basic pattern of valley and mountain winds. Thin lines indicate isobaric surfaces

escarpments, and between extensive highlands and lowlands, such as the Tibetan–Himalayan massif and the Ganges plains in northern India. Obviously, large regional variations of the general pattern occur, but upward winds towards the mountains during the day and downward winds towards the valleys at night prevail in most cases.

Of the two winds, the anabatic winds are usually stronger and more persistent. They frequently continue well after sunset and this tendency is particularly strong in the outer tropics during the summer, when insolation is very intense and nights rather short. Under these circumstances, the anabatic winds, if developed on a large scale, can continue throughout the night. This happens, for instance, at the foothills of the Himalayan mountain range (Flohn, 1960).

Where winds of the general circulation prevail from one direction, as is the case with the trade winds or the monsoons in some areas, anabatic winds generally reinforce the prevailing wind on the windward side of mountains. Here, they can contribute to the orographic rainfall and these areas frequently exhibit a clear afternoon rainfall maximum. However, on leeward slopes the anabatic winds are usually suppressed by the winds of the general circulation.

Katabatic winds are normally weaker than the day time local winds, because thermal differences are usually smaller and friction reduces wind speeds near the earth's surface. The main visible effect of the mountain wind is the rapid dissolution of clouds near the mountain tops or over slopes. In the later hours of

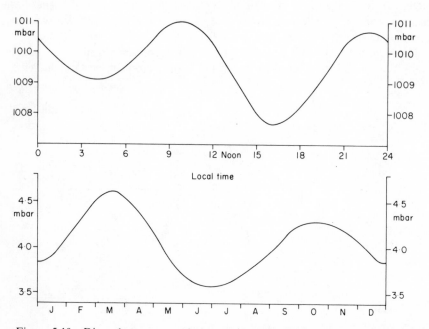

Figure 5.10 Diurnal pressure variation at Singapore Top: hourly variation (annual means). Bottom: seasonal variation of the semi-diurnal amplitude. (Both graphs are based on observations over 18 years)

the night the cool descending air frequently causes fogs in valleys and basins, as continued cooling has increased the humidity to reach saturation. These early morning fogs usually disappear soon after sunrise.

Diurnal pressure variations

An interesting feature of the low latitudes is a quite regular semi-diurnal cycle in air pressure, which shows maxima, of different intensity, around 10.00 and 22.00 hours, and minima at 04.00 and 16.00 hours local time (Figure 5.10). Similar pressure cycles occur at all latitudes, but their amplitudes decrease with distance from the equator, where it reaches 3–4 mbar, to about 1 mbar in the mid-latitudes. Moreover, in the higher latitudes irregular pressure variations, caused by disturbances, fronts and moving weather systems generally predominate strongly in the diurnal march of air pressure, concealing the much smaller regular cycle. In the low latitudes, where irregular changes of pressure are generally smaller and less frequent, the semi-diurnal cycle must be taken into consideration in short-term forecasting and in the calibration altimeters.

The origin of the semi-diurnal cycle is still rather controversial. Resonance oscillations in the atmosphere, diurnal temperature variations and tidal effects have been mentioned as possible causes (Frost, 1960). The seasonal variation of the cycle's amplitude near the equator, which shows maxima around the equinoxes, seems to confirm the relation with direct absorption of solar radiation, probably in the upper levels of the atmosphere (Figure 5.10).

The climatic effects of the semi-diurnal pressure cycle are limited. It has now been established that slight maxima of cloudiness and rainfall around sunset and sunrise are related to it (Brier and Simpson, 1969).

References

Barry, R. G. and Chorley, R. J., 1971, *Atmosphere, Weather and Climate*, 2nd ed. London, Methuen, 379 pp.

Barton, T. F., 1962, Thailand's rainfall distribution by geographic regions, *Journal of Geography*, **51**, 115–116.

Brier, G. W. and Simpson, J., 1969, Tropical cloudiness and rainfall related to pressure and tidal variations, *Quarterly Journal of the Royal Meteorological Society*, **195**, 121–147.

Byers, H. R. and Rodebush, H. R., 1948, Causes of thunderstorms of the Florida peninsula, *Journal of Meteorology*, **5**, 275–280.

Das, P. K., 1972, *The Monsoons*, London, Arnold, 162 pp.

Flohn, H., 1955, *Tropical Circulation Patterns*, Technical Note No. 9, WMO, Geneva, pp. 7–9.

Flohn, H., 1960, Recent investigations on the mechanism of the summer monsoon of southern and eastern Asia, in *Symposium on Monsoons of the World*, New Delhi, 1960, pp. 75–88.

Flohn, H., 1964, The Tropical Easterly Jet and other regional anomalies of the tropical circulation, in *Proceedings of the Symposium on Tropical Meteorology*, Rotorua, N.Z., 1963, Wellington, N.Z., pp. 160–172.

Frost, R., 1960, *Pressure variation over Malaya and the resonance theory*, Scientific Paper No. 4, Air Ministry, Meteorological Office, London, 13 pp.

Jen-Hu Chang, 1967, The Indian summer monsoon, *Geographical Review*, **57**, 373–396.

John, I. G., 1950, *The properties of the upper air over Singapore*, Memoir No. 4, Malayan Meteorological Service, Singapore, pp. 7, 26.

Kendrew, W. G., 1937, *The climates of the continents*, 3rd ed. Oxford, Clarendon Press, pp. 108–109.

Khromov, S. P., 1957, Die geographische Verbreitung der Monsune, *Petermanns Geogr. Mitteilungen*, **101**, 234–237.

Kimble, G. H. T., 1946, Tropical land and sea breezes, with special reference to the East Indies, *Bull. American Met. Society*, **27**, 99–113.

Koteswaram, P., 1958, The easterly jet stream in the tropics, *Tellus*, **10**, 43–57.

Lautensach, H., 1950, Der hochsommerliche Monsun in Süd-und Ostasien, *Peterm. Geogr. Mitteilungen*, **94**, 18–24.

Leroux, M., 1973, Les principales discontinuités africaines: F.I.T.–C.I.O., in *La structure continue de l'équateur météoroligique sur l'Afrique intertropicale*, Dakar, Direction de l'Exploitation Météorologique, pp. 21–36.

Lockwood, J. G., 1965, The Indian monsoon — A review, *Weather*, **20**, 2–8.

Nieuwolt, S., 1968, Diurnal Rainfall Variation in Malaya, *Annals Association of American Geographers*, **58**, 314, 318, 319.

Nieuwolt, S., 1973, Breezes along the Tanzanian East Coast, *Archiv für Meteoroligie, Geophysik u. Bioklimatologie*, Series B, **21**, 189–206.

Pédelaborde, P., 1958, *Les Moussons*, Paris, Armand Colin, 208 pp.

Ramage, C. S., 1964, Diurnal variation of summer rainfall of Malaya, *Journal of Tropical Geography*, **19**, 66.

Ramage, C. S., 1971, *Monsoon Meteorology*, New York and London, Academic, pp. 4–6.

Riehl, H., 1954, *Tropical Meteorology*, New York, McGraw-Hill, pp. 103–105. –

Watts, I. E. M., 1955, *Equatorial Weather, with particular reference to Southeast Asia*, London, University of London Press, pp. 158–159.

Wexler, R., 1946, Theory and observations of land and sea breezes, *Bull. American Met. Society*, **27**, 275–276.

Yin, M. T., 1949, A synoptic–aerologic study of the onset of the summer monsoon over India and Burma, *Journal of Meteorology*, **6**, 393–400.

CHAPTER 6

Tropical Disturbances

The general circulation and pressure distribution of the tropical atmosphere, with their more or less regular variations, described in the previous two chapters, form a background upon which are superimposed numerous temporary systems of low pressure, which are of widely different duration, size and intensity. These 'disturbances', also called 'depressions' or 'cyclones', largely control weather conditions over the areas where they occur, and they are therefore primarily of interest to the meteorologist and forecaster. They are of climatological importance only where they are so frequent or intensive that they have a clear impact on average or extreme conditions. This is the case in many parts of the tropics, because, despite their general irregularity, many tropical disturbances exhibit clear maxima in their frequencies of occurrence, with regard both to their location and to their seasonal and diurnal distribution in time.

In comparison with disturbances of the higher latitudes, those of the tropics show two major differences. The first is the almost complete absence of the fronts so typical of the mid-latitude depressions. In the tropics, most disturbances develop within one air mass. But even when they are related to an air mass boundary, the differences in temperature between two tropical air masses are usually so small that no clear fronts develop. Often an important air mass boundary can be recognized only because it causes numerous disturbances. Therefore, tropical disturbances do not bring the rapid and severe temperature changes associated with fronts, and their climatological importance is caused by variations in rainfall rather than in temperature.

The second difference between tropical and mid-latitude depressions is related to the lower strength of the Coriolis force in areas close to the equator. In the low latitudes, a centre of low pressure is rapidly filled up by compensatory air currents when its cause ceases to be effective. Further away from the equator, this process of equalization is often retarded by the deviation of the surface air currents as a result of the Coriolis force. Therefore, tropical disturbances, and especially those in the very low latitudes, are generally of shorter duration and of a lower intensity than those of the higher latitudes.

The main condition for the formation of tropical disturbances is the presence of warm and humid air masses, in which no inversion exists. This type of air mass can easily become potentially or conditionally unstable. In this state any upward movement of air will be reinforced by the release of large amounts of latent heat of condensation, which is the main source of energy of tropical disturbances. The rising air movement can therefore reach high levels. It causes a centre of low pressure near the earth's surface, which forms the core of the tropical disturbance. This requirement of instability explains why tropical disturbances are less numerous in dry air masses than in humid ones, and why they hardly ever develop over cool ocean currents and areas where the trade wind inversion is strong.

There are several factors which can create instability in suitable air masses and thereby cause tropical disturbances. The most common one is simple *heating from the earth's surface*, which increases the lapse rate of the lower layers of the troposphere. The resulting upward movement of air is called 'convection', and it takes place predominantly in relatively small cells, with a diameter of the order of a few miles. These convection cells are easily identifiable by their typical cumulus or cumulonimbus clouds. Their location can often be explained by surface features which cause locally increased heating, for example, islands, clearings in forested areas, dark coloured plots, slopes exposed to the sun, or bush and forest fires. In many cases, however, the relation is not obvious and identification of the main origin of the convection cells is complicated by the erratic movements of these cells after their formation. In this case the expression 'random distribution of local convection cells' indicates really a lack of knowledge (Gregory, personal communication).

Convection cells frequently turn into *thunderstorms*, the smallest and most frequent form of tropical disturbance. However, if the air masses are relatively dry, or if they are stable because an inversion is present at higher levels, even strong surface heating will not result in sustained upward movement of air. Rather, the rising movement will remain limited to relatively shallow surface layers. If this process takes place over large areas, *heat lows* of considerable size are formed by the expansion of these air layers. These lows are known from desert areas, like the eastern Sudan and the Thar Desert, and they can persist for long periods. They generally weaken at night, when surface temperatures decrease. Heat lows are normally not accompanied by dense clouds, as the upward movement often stops below or near the condensation level of the dry air. If clouds develop, they are usually of the fair weather cumulus type and may cause a weakening of the heat low, as they reduce surface heating. Despite their persistence, heat lows therefore rarely produce any rainfall (Ramage, 1971, pp. 33–36).

Lapse rates can also be increased by *cooling at higher levels*. Apart from advection of cold air masses aloft, which is rare in the tropics, this process is usually the result of radiation losses to outer space. When a cloud cover is present, much of the long wave radiation emanates from the cloud tops, reducing the temperatures in the higher parts of the troposphere during the night. Over large water bodies, where surface temperatures remain practically unchanged during the night, the lapse rate is increased by this fall of temperature aloft, and often instability is the result. The process, called 'nighttime convection', causes thunderstorms over sea and coastal areas during the night or early hours of the morning.

Another group of factors causes instability because they initiate upward air movements. Once started, these movements are often sufficient to produce convectional instability, especially when the lower air layers are very humid, as is frequently the case in the tropics with their large areas of oceans. One factor of this group is *convergence*. When two air masses meet on convergent courses, there is no way for the air to go but upward. Where the air masses differ in temperature, the warmer air rises over the colder one, but at most air mass boundaries in the tropics the temperature differences are very small and upward movements occur on both sides of the zone of convergence and in both air masses (Watts, 1955a, pp. 126–127). The confluence zones are therefore usually accompanied by thunderstorms, organized in bands or '*linear systems*'. These can be hundreds of kilometres long near important zones of convergence, but they are much smaller when the convergence is caused by local winds.

Convergence at lower levels can also take place within one air mass, when the air stream is reduced in speed or when its direction is changed, as was indicated about the trade winds when they reach equatorial latitudes (Chapter 4).

Just as heating at lower levels has a counterpart in cooling at higher levels, the same effects caused by convergence in the lower parts of the troposphere can also be the result of *divergence aloft*. An upper tropospheric high pressure area, which creates a general outflow of air, thereby causes an upward movement of air, which in suitable air masses can result in instability at lower levels. Disturbances are, of course, always connected with some outflow of air aloft, but in some cases the upper air divergence is not the result of the disturbance, but rather the cause of it.

A third group of factors that can start upward movements of air is related to *features of the earth's surface*. Obvious examples are orographic lifting by mountains, hills and coastlines when they form an obstacle in the path of an air stream. But uplifting can also be caused by differences in surface friction, for instance when an air stream moves parallel to a coastline. The stronger friction over the land reduces the effects of the Coriolis force, resulting in convergence and uplifting (Figure 6.1). Friction can cause a slowing down in an airstream or a change in its direction, both of which can result in convergence. Relief features can also create convergences when air masses are moving through a narrow gap between mountain ranges or into a funnel-shaped valley.

The origin of some tropical disturbances lies *outside the tropics*. Cyclones from the Polar Front region occasionally venture far toward the equator, and they regenerate when meeting warm and humid air masses. They tend to lose their frontal characteristics and become true tropical disturbances (Lumb, 1966). Other tropical disturbances are caused by waves in the upper westerlies, from which cold air masses are sheared off towards the equator. The upper air divergence related to this cold core can create *subtropical cyclones* at lower levels. Disturbances of this type occur mainly in the outer tropics, such as over the Pacific Ocean to the north-east of Hawaii, western India and the north-eastern parts of the Arabian Sea (Ramage, 1971, pp. 47–60).

Most tropical disturbances owe their formation to a number of factors,

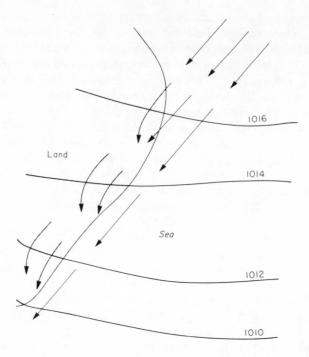

Figure 6.1   Convergence in an air stream parallel to a coastline (northern hemisphere)

though usually one of these can be recognized as the major one. Sometimes even experienced tropical meteorologists are unable to explain the origin and development of disturbances, because over large parts of the tropics observing meteorological stations are spread too thinly to obtain an accurate idea of the conditions under which they were formed. A definite classification of tropical disturbances according to their origin is therefore still impossible; moreover, there exist many local variations of the main types (Krishnamurti, 1972). The following description is limited to a few types, outstanding in their climatological importance because of their high frequency, strong intensity or large size. These are: thunderstorms, monsoon depressions, linear systems, easterly waves and tropical cyclones.

Thunderstorms

It is open to question whether thunderstorms can be properly classified among disturbances, not only because they are of small size but also because of their short duration. They make, therefore, only minor and purely temporary impressions on the general pressure distribution and wind pattern. They can be considered as manifestations of large-scale conditions, rather than factors which make for change in them.

Nevertheless, thunderstorms are of considerable climatological importance in the tropics. They are an elementary unit of the larger tropical disturbances and occur so frequently that they produce a large proportion of the rainfall in most tropical areas. They are therefore included in this chapter.

Thunderstorms are always purely local affairs, as they rarely reach diameters of more than about 10 km. Their duration is limited to 1–2 hours only and, where thunderstorm activity prevails over longer periods, it is merely a repetition of the short process. Because of their limited proportions, thunderstorms are mainly of interest to the meteorological forecaster and descriptions of the physical processes taking place are found in textbooks of meteorology (Byers, 1959; Petterssen, 1969).

It has been estimated that at any one time about 3000 thunderstorms are taking place near the earth's surface. Most of these occur in the tropics, where their intensities are also generally higher than in the mid-latitudes (Blüthgen, 1966, pp. 305–309). Many parts of the tropics have an average of well over 50 days with thunderstorms per year, values which are rarely reached outside the tropics. The map (Figure 6.2) shows clearly the basic conditions for the development of thunderstorms: warm and humid air masses, which can become unstable over considerable vertical layers. Areas where cool or dry air masses prevail, like the eastern parts of the ocean basins and the deserts, have few thunderstorms.

It is generally estimated that a thunderstorm will develop when the instability reaches levels around 8000 m. Whether this condition will be met depends largely on local conditions, which trigger off the process by causing an initial upward air movement. The most common factor is, of course, convection, resulting in widely scattered thunderstorms developing over places where surface heating is most intense. Nighttime convection over water bodies and advection of cold air aloft are other factors which can cause scattered thunderstorms.

Thunderstorms distributed in wide bands or lines are usually the result of orographic lifting along mountain ranges, coastlines or the sea breeze front. They can also be caused by convergence, for instance between sea breezes over an island or peninsula. When these lines of thunderstorms are relatively continuous, they are often called 'squall-lines'. There exists a gradual transition to the 'linear systems', which also consist mainly of thunderstorms but show some organization on a larger scale.

The time of occurrence of thunderstorms depends mainly on the factor which initiates the process. Convectional (or 'thermal') thunderstorms, and those caused by sea breezes and associated phenomena, show a clear maximum during the afternoon. Thunderstorms over lakes and seas, or those caused by land breezes, occur mainly during the night or early hours of the morning. The other categories show no clear diurnal rhythm. Seasonally, thunderstorms show frequencies of occurrence which are closely related to the properties of the prevailing air masses: they are therefore generally rare during seasons when stable or dry air masses prevail over an area.

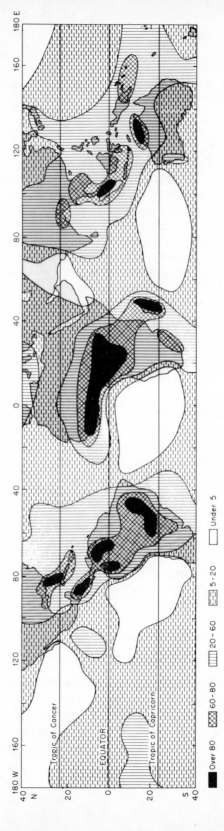

Figure 6.2 Mean number of days with thunderstorms per year

The thunderstorm process itself can often be divided into three stages, each lasting between 20 and 40 minutes. In the beginning stage, strong updrafts prevail in the thunderstorm cell and the cumulus cloud grows rapidly upward, reaching levels around 8000 m. In this stage little or no precipitation takes place, and lightning is rare.

In the second, or mature stage, the thunderstorm reaches it highest intensity. Updraughts continue to be strong, but in some parts of the cell they are replaced by downdraughts. These are accompanied by precipitation. This precipitation is rather intense and strongly localized: it is very high over a small, well defined area, but outside that area no rainfall is received. In this stage, the cumulonimbus cloud frequently reaches levels up to 18 000 m and it often develops an anvil-head, caused by upper tropospheric winds (Figure 6.3). This cloud type is so common in the humid tropics that it is usually associated in the popular imagination with the low latitudes.

In the third, or dissipating stage, downdraughts prevail in most of the cell, and lighter precipitation falls over most of the area covered by the cell.

The climatological importance of thunderstorms is mainly related to the precipitation which they bring. Since most of this falls during the second stage of the process, it is usually characterized by high intensity, short duration and strong localization. Rain gauges only a few miles apart can record very different amounts of rainfall when short periods of observation are used (Jackson, 1969).

The total amount of precipitation caused by one thunderstorm depends not only on its size and intensity, but also on its movements during its most active stage. Thermal thunderstorms and those related to orographic lifting largely

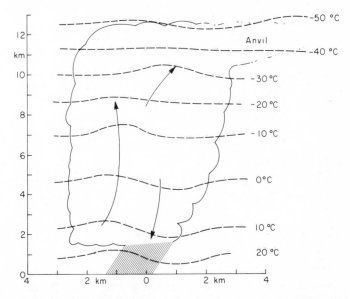

Figure 6.3   A model of a thunderstorm in its mature stage (after Byers, 1959)

develop and decay *in situ*, moving only little during their life cycle. But thunderstorms which develop as the result of convergence generally move with the main air stream in which they are embedded (Sumner, 1971; Watts, 1955b). However, exceptions to this rule are frequent, because many thunderstorms develop as the result of not one but a number of factors. It is often impossible to identify individual thunderstorm cells, as they follow in very close succession. (Orchard and Sumner, 1970).

Because downdraughts can be very strong, some thunderstorm precipitation may be in the form of hail. In the tropics, hail is most common, of course, in the highlands. Still, it has been observed even in equatorial lowlands, as for instance on various occasions in Singapore.

Monsoon depressions

The other types of tropical disturbance have completely different dimensions to thunderstorms, regarding both their size and duration. One type is characterized

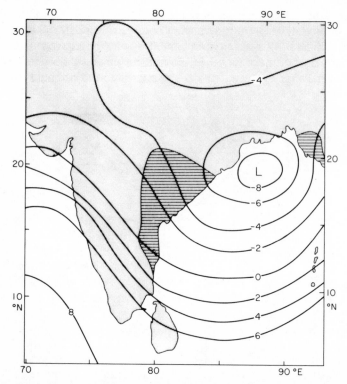

Figure 6.4   An example of a monsoon depression over the Bay of Bengal. 1000-mbar contours in decametres for 12.00 GMT on August 20th, 1967. Hatched areas: continuous rainfall (after Koteswaram and George, 1958)

by quasi-circular or elliptical isobars around a core of low pressure with a diameter in the order of 500–1000 km. The central pressure near the earth's surface is about 3–10 mbar below the normal pressure of the environment. These disturbances last from a few days to about one week, during which time they cause widespread rainfall (Figure 6.4).

Disturbances of this kind occur at the rate of about three per month in the Asian summer monsoon, and they are therefore usually referred to as 'monsoon depressions'. But rather similar disturbances have been observed in the trade winds when they are close to the equator and in the general vicinity of the I.T.C.Z.

In the Asian monsoon, the most important area of origin is the Bay of Bengal, but occasionally monsoon depressions also develop over the Arabian Sea. At both locations their development takes place when the main monsoon trough is very close by. Convergence in the westerly monsoon current near the surface combined with divergent easterlies aloft initiate an upward air movement, which intensifies by the release of large amounts of latent heat. The result is a large low pressure area. The centre of the depression is usually best developed at a level of 2000–4000 m (Ramage, 1971, pp. 43–47).

Once the depression has been formed, it moves slowly WNW with the upper air flow. Over land these depressions usually weaken rapidly. They almost never develop into strong tropical cyclones, mainly because the strong wind shear of the summer monsoon prevents the formation of a vortex.

The climatological importance of these depressions is mainly related to the local intensification of rainfall which they cause. The Indian monsoon depressions bring most rainfall in their southwestern quadrant (Figure 6.4). Because of their irregular occurrence, they bring a certain variability in the monsoon rainfall over the coastal areas of the Bay of Bengal.

Linear systems

Linear systems consist mainly of numerous thunderstorms, organized in lines or bands, which, because of a common origin, develop and move more or less as an organized system. They are also called 'squall lines'. These systems can be hundreds of kilometres long, while their width is usually of the order of 10–30 km. They normally do not consist of uninterrupted cumulonimbus clouds, but exhibit zones of stronger and weaker thunderstorm activity.

Despite many regional and seasonal variations, a definite type of disturbance can be recognized. Their origin is generally due to a combination of different factors, but convergence and convection are the major ones. The latter factor explains why some linear systems are most intense during the afternoons.

Linear systems occur frequently as the result of confluence (and are known as 'disturbance lines' in West Africa (Chapter 5). They produce a considerable, though rather variable, part of the rainfall at the beginning and towards the end of the monsoon season (Gregory, 1965). Similar linear systems occur often during the pre-monsoon season of northwestern India and in Bangladesh.

The confluence can also be caused by local wind systems, and then the linear disturbances are generally much smaller. Convergence of sea breezes is known from many islands, and even from large peninsulas such as Florida and Malaya (Byers and Rodebush, 1948; Watts, 1955a, pp. 124–125). The convergence can also be between land breezes over a narrow strait, such as the Straits of Malacca, or between a local wind and a major air current of the general circulation. The sea breeze especially sometimes develops against the main general wind direction, creating a linear system known as the 'sea breeze front' (Ramage, 1964, pp. 65–66; Nieuwolt, 1973).

As an example of linear systems caused by many factors, the 'sumatras' of southwestern Malaya may be described. They consist of a band of cumulus and cumulonimbus clouds with a length of 200–300 km, which forms over the Straits of Malacca. These disturbances develop only during the south-west monsoon season (May–September), usually during the night, and they move slowly with the main monsoon current towards the coast of Malaya, where they arrive during the late hours of the night or in the early morning. They are most frequent between Singapore and Port Swettenham, where the Straits are only 40–80 km wide, and occur less often further north.

The origin of the sumatras is not related to a major air boundary, because most of them are formed within one air stream. Three main factors can usually be recognized in their development. Firstly, the air masses of the rather sluggish south-west monsoon are heated, during the day, over the extensive lowlands of Sumatra, and they often become convectionally unstable. During the night, when they are situated over the Straits of Malacca, this instability is reinforced by radiation losses from the cloud tops.

Secondly, when those air masses reach the Malayan south-west coast, they are uplifted orographically. They can also be undercut by the land breeze. Though this breeze is usually rather weak when cloudy conditions prevail, it is locally intensified where the form of the coastline is concave, resulting in convergent land breezes.

Thirdly, where the Straits of Malacca are narrow, land breezes from Malaya and Sumatra may converge. This last factor explains why sumatras are most frequent in the southern parts of the Straits (Nieuwolt, 1969; Ramage, 1964, p. 64; Watts, 1955a, pp. 168–173).

The duration of a sumatra is about 1–2 hours. Wind gusts accompanying the squall can reach speeds up to 70 km/h. But the real importance of the sumatras lies in the rainfall which they bring. An individual sumatra can bring up to 80 mm, and because of their high frequency the total proportion of rainfall caused by sumatras is considerable. They are largely responsible for the diurnal rainfall maximum during the early hours of the morning which prevails at Malacca and other stations along the south-west coast of Malaya during the south-west monsoon season (Nieuwolt, 1968).

Easterly waves

Although the trade winds are generally very steady and regular winds, which bring stable weather conditions over most areas where they prevail, in some parts of the tropics this tranquil situation is at times interrupted by disturbances. These disturbances show a number of variations in size and intensity, but they have one characteristic in common: their main centre of low pressure is not circular or elliptic, but in the form of a wave in the isobaric pattern; and the main trough line is at approximately right-angles to the main trade wind direction (Figure 6.5). As these waves move with the easterlies, they are generally called 'easterly waves'. Some of the easterly waves may also contain small vortices, but these are usually of minor importance.

The main areas where easterly waves occur are the western parts of the large ocean basins, in latitudes between about 5 and 20 degrees. They are best known in the Caribbean area, but very similar disturbances have been observed in the Pacific Ocean, north of the equator. Relatively little is known about their occurrence in the southern hemisphere (Riehl, 1954, pp. 210–234; Malkus and Riehl, 1964).

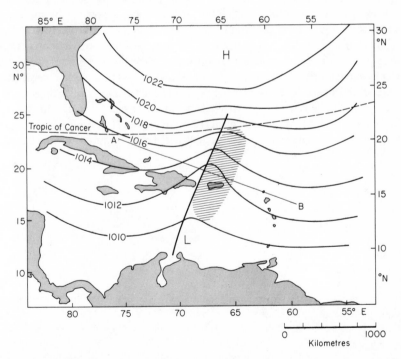

Figure 6.5　Model of an easterly wave in the Caribbean area. Hatched area indicates main rainfall zone

Easterly waves are most frequent during the late summer, when they may come as often as every 3–5 days in the Caribbean area and every 2–3 days near Guam in the Pacific Ocean. This high frequency is related to the weakening of the trade wind inversion during that period, caused by the high surface water temperatures, which reach their annual maximum towards the end of the summer.

The formation of an easterly wave is often triggered off by small vortices, which travel with the trades, sometimes over very long distances. For some easterly waves in the Caribbean these vortices have been traced as far eastwards as West Africa, where they may have originated from the remnants of mid-latitude depressions which had strayed far towards the equator. In other cases a more direct interaction with mid-latitude troughs might have initiated the development, and it is also possible that friction over islands starts the process leading to an easterly wave (Riehl, 1954, pp. 225–226).

A typical easterly wave covers a large area (Figure 6.5). It is usually relatively weak near the surface, reaching its strongest development at levels around 4000 m (Figure 6.6). Easterly waves generally move with speeds of about 15–20 km/h. At lower levels, where the trades are strong, they therefore lag behind the main wind; but at higher levels, over about 2000 m, where the trades are weaker, the easterly waves move ahead of them. It can be calculated that under these flow conditions a divergence at low levels is overlain by high level convergence, while behind the main wave axis the opposite movement results (Riehl, 1954, pp. 219–223). Weather conditions associated with easterly waves certainly confirm these theoretical conclusions: ahead of the easterly wave the weather is particularly fair, as the trade wind inversion is temporarily reinforced and moved to a lower level. But at the rear of the main axis, squall lines prevail and the trade wind inversion is destroyed. It usually re-establishes itself, but at a much higher level than before the easterly wave passed (Figure 6.6). The squall lines often show a pattern more or less parallel to the main wind direction.

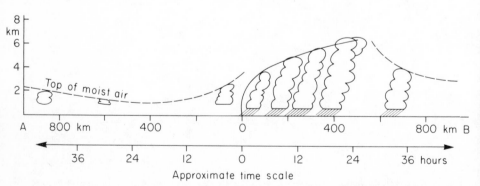

Figure 6.6 Cross-section along the line A–B in Figure 6.5, showing the vertical structure of an easterly wave. Vertical exaggeration about 50 times

Easterly waves are climatologically important because they bring large amounts of rainfall to areas which are generally dry as long as the trades are undisturbed. Easterly waves are the main reason for the late summer rainfall maximum observed in many islands of the Caribbean and in the western parts of the Pacific Ocean. In these areas the rainfall is strongly concentrated during only a few days per year. For instance at Oahu (Hawaiian Islands), over a period of 10 years, two-thirds of the annual total rainfall was normally received during about 10 days, when easterly waves moved over (Mink, 1960). Easterly waves are also of great interest because they occasionally develop into tropical cyclones.

Tropical cyclones

Strong tropical cyclones are called 'hurricanes' in the Caribbean area, 'typhoons' in the Far East, and 'willy-willies' in Australia, but despite some regional differences they are all basically of the same type. They are characterized by almost circular isobars around a centre of very low pressure, normally around 950 mbar, which occasionally can reach values as low as 920 mbar at sea level. The systems have a diameter of around 500–800 km, but some typhoons have been much larger. Because very strong pressure gradients

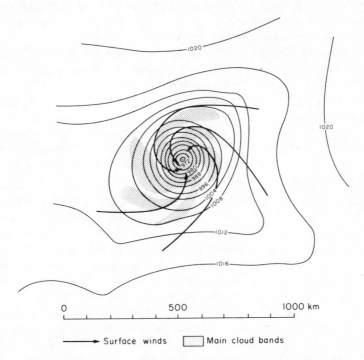

Figure 6.7   Model of a tropical cyclone. Thin lines represent sea-level isobars

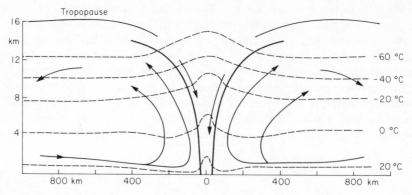

Figure 6.8   Cross-section of a tropical cyclone. Vertical exaggeration about 50 times

prevail near the centre of the cyclone, wind speeds can be high: they usually exceed 120 km/h and winds of over 200 km/m have been observed. In many cases higher wind speeds could not be recorded because instruments had been damaged or destroyed. Satellite pictures of tropical cyclones reveal a cloud pattern in spiral bands around the centre of the depression (Figure 6.7). These contain a large number of cumulonimbus towers, which may reach heights of 10 to 12 km. A typical cyclone includes around 100–200 of these, but numbers vary a great deal according to the size and intensity of the individual tropical cyclone.

The centre of the system is formed by a circular area with a diameter of about 10–40 km, in which subsiding air and calms prevail, and cloudiness is light. This is the 'eye' of the storm, a typical feature of tropical cyclones (Figure 6.7).

The inflow of air towards the centre of the depression is about 1000–2000 m deep, while the outflow takes place in the upper troposphere at levels between 10 000 and 15 000 m (Figure 6.8).

Because tropical cyclones can do extensive damage, they have been studied extensively, mainly in the Caribbean area and in the Pacific Ocean (Palmén and Newton, 1969, pp. 471–522; Riehl, 1954, pp. 281–357). It has been established that five main conditions are necessary for their formation:

(1) The rising air masses in the core of the depression must be warmer than the surrounding air masses up to a level of about 10 000–12 000 m. As the main driving force of tropical cyclones is latent heat of condensation, the rising air masses must also be rather humid. These disturbances will therefore form only over large ocean areas, where the water surface temperature is above the critical value of 27 °C (Figure 6.9). This condition explains why tropical cyclones originate mainly over the western parts of the large ocean basins where no cold currents occur. It also accounts for the fact that the main season of the tropical cyclones is towards the end of the summer, when sea surface temperatures are highest.

(2) Unlike the situation in other tropical disturbances, vorticity is an essential part of the circulation in tropical cyclones. Therefore they do not develop at

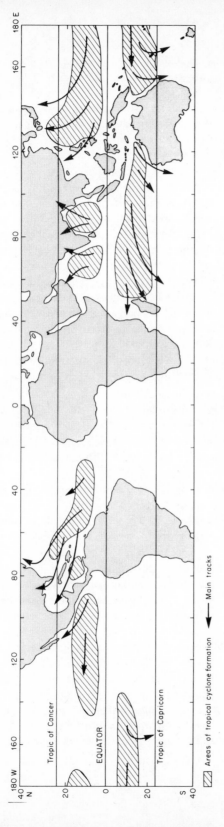

Figure 6.9 Formation and movements of tropical cyclones

▨ Areas of tropical cyclone formation ⟶ Main tracks

latitudes below about 5 degrees. Closer to the equator, the Coriolis force is too weak to divert the inflowing air streams and even a strong surface low will fill up rapidly.

(3) The basic air current in which the tropical cyclone is formed should have only weak vertical wind shear, since vertical shear inhibits the development of a vortex. This is the main reason why no tropical cyclones develop in the Asian summer monsoon when it is at full strength. Both over theBay of Bengal and in the Arabian Sea tropical cyclones have their highest frequencies at the beginning and towards the end of the summer monsoon season (Blüthgen 1966, p. 303).

(4) A small low pressure centre is required to trigger off the whole cyclone development. The initiative can come from small vortices near the I.T.C.Z. or from easterly waves. This condition explains why tropical cyclones are almost unknown in the southern Atlantic Ocean and the southeastern Pacific, where the I.T.C.Z. is generally close to the equator or in the other hemisphere (Figure 4.2). However, in many cases it is not possible to trace the original disturbance which started the development of a cyclone and it is therefore possible that other mechanisms occasionally start their formation.

(5) Combined with a surface low, an area of relatively high pressure should be present at levels between 9000 m and 15 000 m, so that the outflow of air in the upper troposphere is maintained (Riehl, 1954, pp. 300–307). This condition explains why many incipient cyclones fail to develop in intensity.

Since cyclogenesis takes place over ocean areas, the actual atmospheric conditions at the time of formation are frequently not known, and therefore some other factors might be involved. In some cases, small vortices or easterly waves do not develop into tropical cyclones, though all the above conditions seem to be met.

Once they are formed, the tropical cyclones move with the upper tropospheric winds. Over the oceans, this means that they move westwards, and their speed is usually of the order of 15–30 km/h. They seem to follow certain preferred tracks, which are related to warm ocean currents (Figure 6.9). When they reach the western margins of the ocean basins, the tropical cyclones 'recurve' polewards or move into the continents. Either way they usually decay rapidly, because the necessary supply of large masses of very warm and humid air is limited. The decay over land may also be caused by increased surface friction, lower temperatures especially during the night, and increased vertical wind shear.

In the Bay of Bengal and the Arabian Sea the tropical cyclones usually move towards the continent soon after their formation. Their life spans are therefore shorter than those of the oceanic cyclones. However, both in their movements and in their development cyclones are notoriously erratic and the prediction of their tracks is one of the main problems in tropical forecasting.

The frequency of occurrence of tropical cyclones exhibits large variations from year to year. The mean frequency per 10 years in the various ocean basins is:

North-west Pacific Ocean	208
North Atlantic Ocean, Caribbean	85
Bay of Bengal	75
South-west Indian Ocean	41
North-east of Australia	31
North-west of Australia	23
Arabian Sea	19
North-east Pacific Ocean	10

During the last few decades, there seems to have been a tendency towards higher frequencies, but it is uncertain whether this trend will continue or is part of a cyclic change (Milton, 1974).

Tropical cyclones have a life span of about one week. During this period various phases can often be recognized, but individual cyclones frequently deviate largely from an established scheme or cycle. Usually the cyclone grows first in intensity, until the central low pressure reaches values of 950–900 mbar and the eye is clearly developed. Then it expands in size, while still maintaining its high intensity. Decay starts over land or cooler water surfaces (Palmén and Newton, 1969, pp. 505–507; Riehl, 1954, pp. 282–284).

Rainfall in tropical cyclones is usually concentrated in a rather narrow zone around the core, but the actual amounts received vary largely, depending on size, intensity and movements of the individual disturbance. In a moving cyclone, rainfall in a circle with a diameter of about 100 km around the centre has been calculated at 863 mm per day (Riehl, 1954, pp. 294–295). How much of this will be received at one station depends, of course, on the time that this zone is situated over the station. The amounts of rainfall can also be increased sharply when orographic lifting takes place. Values of over 2000 mm per day have been recorded in the Philippines under these circumstances. However, because of the extremely high wind speeds exact measurements of rainfall in cyclones are often impossible. Tropical cyclones are the main reason for late summer or autumn rainfall maxima in many parts of the tropics (Chapter 9).

Finally, it should be emphasized that many tropical disturbances do not fit clearly into one of the five groups dealt with in this chapter, because they contain characteristics of more than one category. This may be due to the rather individual development of a disturbance or to local conditions which create strong variations of the main types.

References

Blüthgen, J., 1966, *Allgemeine Klimageographie*, 2nd ed., Berlin, Walter de Gruyter, 720 pp.

Byers, H. R., 1959, *General Meteorology*, New York, McGraw-Hill, pp. 458–475.

Byers, H. R. and Rodebush, H. R., 1948, Causes of thunderstorms of the Florida Peninsula, *Journal of Meteorology*, 5, 275–280

Gregory, S., 1965, *Rainfall over Sierra Leone*, Liverpool, University of Liverpool Department of Geography, No. 2, p. 15.

Jackson, I. J., 1969, Tropical rainfall variations over a small area, *Journal of Hydrology*, **8**, 99–110.

Koteswaram, P. and George, C. A., 1958, A monsoon depression in the Bay of Bengal, in S. Basu *et al.* (eds.), *Monsoons of the World*, New Delhi, Hind Union Press, pp. 154–146.

Krishnamurti, T. N., 1972, Transient disturbances in the tropics, in J. A. Young (ed.), *Dynamics of the tropical Atmosphere*, Boulder, Col., National Center for Atmospheric Research, pp. 45–73.

Lumb, F. E., 1966, Synoptic disturbances causing rainy periods along the East African coast, *Meteorological Magazine*, **95**, 150–159.

Malkus, J. S. and Riehl, H., 1964, *Cloud structure and distributions over the tropical Pacific Ocean*, Los Angeles, California University Press, 229 pp.

Milton, D., 1974, Some observations of global trends in tropical cyclone frequencies, *Weather*, **29**, 267–270.

Mink, J. F., 1960, Distribution pattern of rainfall in the leeward Koolau mountains, Oahu, Hawaii, *Journal of Geophysical Research*, **65**, 2869–2876.

Nieuwolt, S., 1968, Diurnal rainfall variation in Malaya, *Annals, Association of American Geographers*, **58**, 319–322.

Nieuwolt, S., 1969, *Klimageographie der malaiischen Halbinsel*, Mainz, Geogr. Institut der Johannes Gutenberg – Universität, p. 39.

Nieuwolt, S., 1973, Breezes along the Tanzanian east coast, *Archiv für Meteorologie, Geophysik u. Bioklimatologie*, Series B, **21**, 194.

Orchard, A. Q. and Sumner, G. N., 1970, *Network Report No. 4, East African Rainfall Project, King's College*, London, 72 pp.

Palmén, E. and Newton, C. W., 1969, *Atmospheric Circulation Systems, their Structure and Physical Interpretation*, New York, Academic, 603 pp.

Petterssen, S., 1969, *Introduction to Meteorology*, 3rd ed., Tokyo, McGraw-Hill, pp. 120–128.

Ramage, C. S., 1964, Diurnal variation of summer rainfall of Malaya, *Journal of Tropical Geography*, **19**, 62–68.

Ramage, C. S., 1971, *Monsoon Meteorology*, London, Academic, 296 pp.

Riehl, H., 1954, *Tropical Meteorology*, New York, McGraw-Hill, 392 pp.

Sumner, G. N., 1971, *Observations on the weather at Nairobi, Kampala and Dar es Salaam between January and April, 1971*, Special Notes No. 1, East African Rainfall Project, King's College, London, 43 pp.

Watts, I. E. M., 1955a, *Equatorial Weather*, London, University of London Press, 224 pp.

Watts, I. E. M., 1955b, *Rainfall of Singapore Island, Journal of Tropical Geography*, **7**, 43–53.

CHAPTER 7

Water in the Tropical Atmosphere

Water in the atmosphere exists in two different forms: as water vapour and, in clouds, as water in its liquid and solid phases. The water component of the atmosphere is highly variable in relation to both time and place.

The climatological importance of atmospheric water is related not only to its capacity to control precipitation, but also to its strong influence on both reflection and absorption of terrestrial and solar radiation. It therefore has a large effect on atmospheric temperature conditions (Chapter 2).

In the tropics, where the atmospheric water content is comparatively high, its role in climate is especially important. As has been shown in the previous chapter, stability conditions of tropical air masses and the latent heat of condensation are the main driving forces of tropical disturbances, and both factors are closely related to atmospheric moisture.

This chapter follows the atmospheric part of the hydrologic cycle, starting with evapotranspiration, the origin of atmospheric water vapour. Then the humidity of the tropical atmosphere is described, followed by the process of condensation. Lastly the products of condensation — clouds — are dealt with. The remaining stage of the hydrologic cycle, the return of atmospheric water to the earth's surface in the form of precipitation, has a special importance in the tropical climates. It will be the subject of the next chapter.

Evapotranspiration

All atmospheric moisture originates from the earth's surface, where water in its liquid and solid phase is transformed into water vapour, which can be carried upward by air movements. The water vapour originates from two processes. The first is *evaporation*, which takes place from water and ice surfaces, and over land areas when soils, rocks or the vegetation cover are wet after recent precipitation or when the ground water is close to the surface. The second source is *transpiration*, physically the same process as evaporation, but carried out by organisms, mainly plants. The combined effects of these two processes are often called *evapotranspiration*, which indicates the total flow of water vapour into the atmosphere (Thornthwaite, 1948).

Evapotranspiration is controlled by three conditions: the capacity of the air to take up more water vapour, the amount of energy available for the latent heat used in the processes of evaporation and transpiration, and the degree of

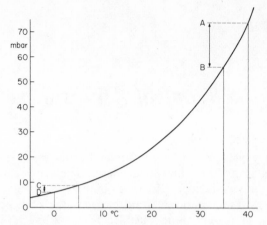

Figure 7.1 Saturation vapoure pressure (in mbar) in relation to temperature. A–B, C–D: amount of water vapour that condenses when saturated air at 40 °C, 5 °C, respectively, is cooled by 5 degrees Centigrade.

turbulence of the lower parts of the atmosphere, necessary to replace the saturated air layers near the earth's surface by unsaturated air from higher levels. In respect of these three requirements the conditions in the tropics are generally rather favourable. In the first place, the capacity of the air to retain water vapour increases rapidly with temperature (Figure 7.1). Warm tropical air masses can therefore take up more water vapour than cold ones. The actual amount that can be absorbed into the atmosphere depends also on the relative humidity of the air — the lower this is, the more favourable are the conditions for further evapotranspiration. The dry tropics have, therefore, very high rates of evapotranspiration.

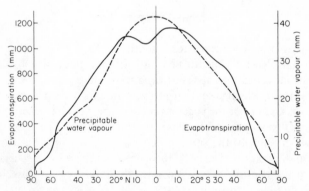

Figure 7.2 Latitudinal variation of mean annual evapotranspiration (solid line) (after Wüst, 1922) and mean precipitable water vapour content of the atmosphere (broken line) (after Sellers, 1965, p. 5). (Precipitable water vapour: total in a vertical column of air of 1 square centimetre from the earth's surface to the top of the atmosphere)

Secondly, the energy for evapotranspiration is mainly provided by solar radiation, which is available in the largest quantities in the tropics. Thirdly, turbulence is caused by winds or by convectional currents and, while winds are not particularly strong in the tropics compared to other climates, convection is very frequent.

It is for these reasons that the latitudinal distribution of evapotranspiration shows a clear maximum in the low latitudes (Figure 7.2). The slightly lower values between the equator and about 10° N are caused by the heavy cloud cover, frequently saturated air masses and low wind velocities related to the I.T.C.Z., which prevails over these latitudes in the oceanic areas during most of the year.

Independent of atmospheric factors, evapotranspiration is often limited by the availability of water. Over water surfaces, where this supply is plentiful, evapotranspiration is not restricted and the values reached are called the 'potential evapotranspiration'. But over land areas water is often a limiting factor and the 'actual evapotranspiration' remains then lower than the potential value. Over large land areas its maximum is set by the total amount of precipitation received. This is the main reason why the oceans have a mean actual evapotranspiration rate of about 930 mm, while the continents reach values of about 500 mm per year (Wüst, 1936). However, in some parts of the humid tropics the continental rate of the actual evapotranspiration may occasionally exceed the potential value recorded over the oceans in the same climatic area. This is caused by the enormous increase of the transpiration surface of the luxuriant rainforests with their multiple canopies (Tison, 1960; Wüst, 1922).

Maps of the mean actual evapotranspiration show that oceanic rates are strongly affected by sea surface water temperatures. Warm oceans reach values over 2000 mm per year but cold oceans rarely have totals of 1000 mm. Over the continents the equatorial areas have the highest rates, with annual totals over 600 mm, but the dry tropics rarely reach half that amount, as evapotranspiration is severely limited by the lack of rainfall (Barry, 1969).

Evaporation pans and other evaporimeters indicate the potential evapotranspiration, as water is always available. However, these instruments often indicate inflated values. This is due to the 'oasis effect', caused by the relatively small evaporating surface. Saturated air layers over the pan are very frequently replaced by relatively dry air from the surrounding areas, where normally no water is available for evaporation. Over large water bodies, such as lakes, or over densely vegetated areas, the replacing air layers come mainly from other parts of the evaporating or transpiring surface. This is the main reason why figures from evaporation pans are often corrected by a factor around 0·7 to make them comparable to evaporation data from lakes and reservoirs (Wiesner, 1970).

Evaporation pans also suffer from other inaccuracies, due to the direct absorption of solar radiation or heat from the ground, or splashing and interference by animals. Therefore the potential evapotranspiration, also called the 'evaporation from an open water surface', is often estimated on the basis of the meteorological factors which control its rate, because these factors can be measured quite accurately (Penman, 1948, 1949; Thornthwaite, 1948). Fortunately, potential evapotranspiration is a conservative element, which shows only small variations from year to year, so that even relatively short records can be used to obtain a reasonably exact estimate of its mean rate per month or per day.

Seasonal variations of potential evapotranspiration are generally small in the tropics, but they show a gradual increase with latitude, mainly because of seasonal differences in solar radiation and temperature (Figure 7.3). It should be borne in mind that the high values at Enugu are possibly caused by the fact that they have been recorded at evaporation pans, while the other data have been computed from the meteorological factors according to the Penman formula. The diagram clearly shows the many irregularities in evapotranspiration caused by local and regional circumstances.

The diurnal variation of potential evapotranspiration is simple: a clear maximum during the day is followed by a very low minimum at night. In the tropics, where the nights are long and the relative humidity near the earth's surface is frequently close to 100%, and where turbulence is limited, the potential evapotranspiration comes to an almost complete standstill at most stations during the night.

Humidity of the tropical atmosphere

The term 'humidity' refers to the content in water vapour of the air. As its source is the earth's surface, atmospheric water vapour is always strongly concentrated in the lower layers and normally about half of its total is found below the level of 2000 m.

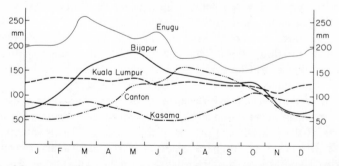

Figure 7.3 Mean potential evapotranspiration at: Kuala Lumpur, Malaysia, 3°7′ N, 101°42′ E; Enugu, Nigeria, 6°28′ N, 7°33′ E (evaporation pan data); Kasama, Zambia, 10°12′ S, 31°11′ E; Bijapur, India, 16°53′ N, 75°42′ E; Canton, China, 23°07′ N, 113°15′ E. Data at all stations except Enugu based on estimates according to the Penman formula

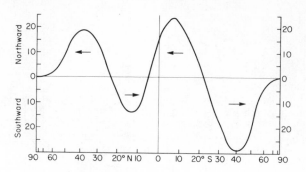

Figure 7.4 The mean annual meridional transfer of water vapour in the atmosphere (in 10^{15} kg) (after Sellers, 1965, p. 94)

The latitudinal distribution of atmospheric humidity shows a very simple pattern, with a clear maximumnear the equator (Figure 7.2). This maximum is, of course, the effect of the high evapotranspiration rates in the low latitudes and the prevalence of warm air masses, which can contain large amounts of water vapour. However, not all the humidity of the tropical atmosphere originates from the same latitude: large meridional transfers of water vapour take place in the tropics, mainly by the trade winds (Figure 7.4). The diagram also indicates that part of the atmospheric humidity of the polar areas has its origin in the outer tropics.

The distribution of the mean atmospheric vapour pressure shows large differences in the tropics (Figure 7.5). Over the oceans, atmospheric humidity is mainly controlled by the sea surface water temperature and it is normally close to the saturation vapour pressure of that temperature. Therefore the western parts of the ocean basins and the equatorial oceans have generally high humidity values. This effect is transferred to the adjacent continents, though in eastern Africa this influence is limited to a narrow coastal strip owing to topographic and dynamic effects. The eastern ocean basins, where cold currents prevail, have much lower vapour pressures.

Over the continents the atmospheric humidity depends on the characteristics of the prevailing air masses. This is clearly demonstrated by the large seasonal differences in humidity in the monsoonal areas of southern Asia and West Africa, and over the continent of South America to the east of the Andes mountain ranges. The dry tropical areas have, of course, low atmospheric humidity throughout the year, but values during the winter of their respective hemispheres are lower than during the summer, as shown by northern and southern Africa and the central parts of Australia. Mountainous areas also have low atmospheric humidity values, because the lower, and more humid, parts of the atmosphere are missing (Figure 7.5).

To illustrate some regional and local variations, Table 7.1 provides mean humidity data. The effects of elevation are shown by the station pair Singapore–Tanah Rata, in equatorial conditions where seasonal differences in

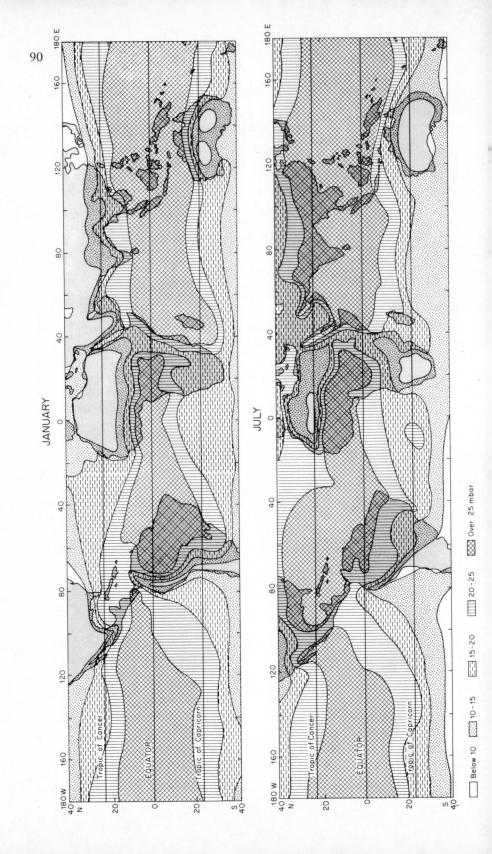

atmospheric humidity are very small, and by the Zimbabwe stations, where the effects are limited during the dry season, but considerable in the rainy part of the year.

Seasonal differences in monsoon climates are illustrated by the two stations in southern and eastern Asia.

The Tanzanian station pair shows the combined effects of elevation and distance from the ocean, Tabora being about 750 km from the Indian Ocean coast where Dar es Salaam is situated. The reduction in mean vapour pressure remains the same in the dry and wet seasons.

The Ugandan station, near Lake Victoria, is compared with a Kenyan station at the same elevation to illustrate the influence of the lake on the humidity conditions. This is, of course, a purely local effect, experienced by many stations near water surfaces.

The two stations in Malagasy are used to indicate that the low level humidity conditions are not affected by their situations on either side of the large mountain ranges of Madagascar. Despite their large differences in rainfall (Tamatave has an annual mean of 3526 mm, while Maintirano receives only 998 mm per year) the two stations differ very little in air humidity.

Finally, two extra-tropical stations are given to illustrate the quite different magnitudes of vapour pressure in comparison with the tropics.

While the seasonal variations in atmospheric humidity follow a simple pattern in most parts of the tropics, being controlled mainly by temperature and air mass characteristics, the diurnal variations are more complicated. At most

Table 7.1   Mean atmospheric vapour pressure (in mbar)

Station	Lat.	Long.	Elevation (m)	January	July
Singapore	1° N	104° E	8	27	30
Tanah Rata, Malaysia	4° N	101° E	1448	17	17
Calcutta, India	23° N	88° E	6	15	34
Koshun, Taiwan	23° N	120° E	10	18	31
Dar es Salaam, Tanzania	7° S	39° E	14	30	23
Tabora, Tanzania	5° S	33° E	1190	20	13
Entebbe, Uganda	0	32° E	1146	20	20
Moyale, Kenya	4° N	39° E	1113	15	16
Bulawayo, Zimbabwe	20° S	29° E	1344	17	9
Beitbridge, Zimbabwe	22° S	30° E	456	22	10
Tamatave, Malagasy	18° S	49° E	5	29	21
Maintirano, Malagasy	18° S	44° E	23	30	20
London, England	52° N	0	33	7	14
Cairo, Egypt	30° N	31° E	30	8	17

tropical stations, the daily minimum of vapour pressure is recorded at night, shortly before sunrise. Temperatures are then at a minimum, so that the lower air layers frequently get saturated and dew is formed, which abstracts water vapour from the air. During the early morning hours, most of the dew is readily available for evaporation and transpiration is high, since the plants have abundant water available; a maximum of vapour pressure builds up. Later in the morning, when the dew has disappeared and transpiration and evaporation are limited by water supply, and when convectional currents develop, which transfer water vapour to higher elevations, the atmospheric humidity decreases to a secondary minimum. Unless precipitation occurs, this minimum persists throughout the afternoon. After sunset, when the temperatures decrease and convectional activity dies down, atmospheric humidity increases again, and this increase is particularly strong after rainfall (Figure 7.6). The diagram for Singapore, which is based on annual means, as length of day hardly varies so close to the equator, illustrates the small decrease during the afternoon, caused by frequent rainfall in this part of the day. The Lusaka diagrams, separated for January and July to allow them to show the difference in length of day (earlier morning and later evening maxima in July), also illustrate the differences between rainy and dry seasons. January in Lusaka, the middle of the rainy season, brings not only much higher vapour pressure values, but also a modest afternoon minimum. In July the minimum is better developed and persists for a longer period. The very clear morning maximum illustrates the importance of dew formation during the rather cold nights of winter.

Over oceans, however, the vapour pressure shows almost no diurnal variations, because the sea surface water temperatures remain practically unchanged.

Relative humidity

In many meteorological publications the moisture content of the air is indicated by the relative humidity, because it has a strong influence on evaporation and transpiration and the probability of convectional rain.

However, relative humidity figures have the disadvantage that they depend very much on the accompanying temperature. In the tropics, where diurnal temperature variations are generally large, the relative humidity therefore changes considerably in the course of the day. At night, the relative humidity frequently reaches values close to 100 per cent at many stations in the tropics (Figure 7.7(a)). For the comparison of different stations, relative humidity data can therefore only be used when they have been observed at approximately the same hour and when the temperatures are not too different. Unfortunately, international practices in the observation of relative humidity vary considerably and no standard hours of readings exist.

Where comparisons are possible, relative humidity values show a clear maximum near the equator, where high rates of evapotranspiration, large

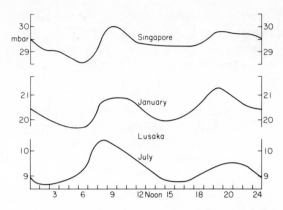

Figure 7.6 Diurnal variation of mean vapour pressure at: Singapore (1°21′ N, 103°54′ E, 8 m), annual means, and Lusaka (15°19′ S, 28°15′ E, 1154 m) for January and July

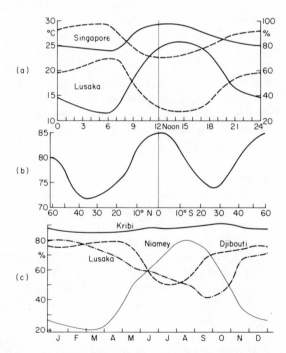

Figure 7.7 Relative humidity. (a) Diurnal variation of temperature (full lines) and relative humidity (broken lines) at Singapore (annual values) and Lusaka (August). (b) Latitudinal distribution of mean annual relative humidity. (c) Monthly mean relative humidity at: Kribi, Cameroons, 2°57′ N, 9°54′ E, 624 m ((max. + min.)/2); Djibouti, A. & I., 11°31′ N, 43°09′ E, 7 m ((6.00 + 12.00 hrs.)/2); Niamey, Niger, 13°30′ N, 2°07′ E, 220 m ((6.00 + 12.00 hrs)/2); Lusaka, Zambia, 15°19′ S, 28°27′ E, 1154 m ((6.00 + 14.00 hrs.)/2).

amounts of advected water vapour and frequent precipitation keep the air humid at all seasons (Figure 7.7(b)). There is a general decrease towards the latitudes of the subtropical high pressure cells, where adiabatic heating of subsiding air masses keep the relative humidity rather low and well away from saturation.

Seasonal variations of relative humidity show a similar pattern to those of vapour pressure, with small variations near the equator, increased differences with latitude and many regional and local variations (Figure 7.7(c)). It is generally better to use the more conservative absolute indicators of atmospheric moisture, such as vapour pressure or mixing ratio, to illustrate climatic conditions.

Condensation

In the atmosphere, condensation takes place when air is cooled beyond its dewpoint. This cooling can be the result of radiation losses during the night and this process causes the formation of dew or morning fog. But more important and widespread is the cooling of air by expansion, when it moves upward and reaches levels of lower pressure. This form of cooling results in clouds.

There is a quantitative difference in the amount of liquid water produced by these processes between the tropical and extra-tropical climates, because warm air masses change their capacity to retain water vapour much faster than cooler ones (Figure 7.1). Actual amounts of water produced by the cooling of one cubic metre of air by five degrees Centigrade decrease from 11·5° grams for air starting at 40 °C, through 4·5 grams for air at 20 °C to only 2·0 grams at 5 °C.

Therefore warm air masses generally produce more water by condensation than cooler ones. The actual level where condensation starts depends, of course, on the relative humidity of the rising air. It is generally higher in warm air masses. But above the rising condensation level, latent heat set free maintains the buoyancy of warm air masses over larger vertical layers than in cold air. Therefore, in the tropics, condensation is not only more vigorous, but also extends over thicker layers of air than in cooler climates.

Dew and morning fog are frequent occurrences in many parts of the tropics. During rainy periods, the humidity of the air at night close to the earth's surface is usually near the saturation point, so that only a slight amount of cooling is sufficient to produce dew. In dry periods, when the humidity of the air at night is much lower, cooling during clear nights is much stronger and even then dew will form quite often.

Morning fog results from the cooling of thicker layers of air. It is therefore usually related to air drainage and morning fogs occur mainly in basins and valleys. Near coasts, where land breezes often produce some turbulence during the night, morning fogs are rare (Watts, 1955, pp. 35–36). Morning fogs disappear soon after sunrise, as a rule, as solar heating and increased turbulence cause warming of the lowest air layers.

Fogs caused by the cooling of moist air over cold ocean waters are particularly frequent in dry tropical areas on the western coasts of the continents, as in Namibia, Morocco, northern Chile and Peru. These fogs often persist for long periods, as daytime heating by insolation is prevented from reaching the earth's surface by them and remains inefficient over ocean surfaces in any case.

Tropical clouds

The main products of condensation in the atmosphere are clouds. Clouds are climatologically significant, not only because they are the source of precipitation and control the intensity of solar radiation during the day and the extent of radiation losses at night, but also because they are the main controlling factor of the otherwise rather uniform temperatures in many parts of the tropics (Nieuwolt, 1968). Not only are cloudy days cooler than sunny ones, and cloudy nights generally warmer than clear nights, but clouds can also cause large short-term variations in temperature. A rapidly developing cumulus cloud can produce a fall of temperature of the order of five degrees Centigrade within one hour, before the thunderstorm associated with it actually starts (Nieuwolt, 1966).

Clouds strongly influence subjective impressions of climatic conditions. Dull and overcast days tend to have a depressive effect on most people, while bright days create a more positive feeling. This is particularly the case when a few scattered clouds accentuate the blueness of the sky. In this respect the tropics are favoured, because overcast days are generally rare. Cumulus clouds, developing with great speed, are an impressive and exciting sight, often associated with tropical afternoons. Only a few tropical coastal areas along cold ocean currents have more frequent heavy cloudiness.

At night the psychological impact of clouds is much smaller. Nevertheless, the bright starlit evenings and nights are one of the attractions of the tropical climates, and not only because they bring relief from the heat of the day.

In the tropics, the various cloud types occur at higher levels than in the mid-latitudes. This is especially true for high and medium level clouds, which are related to sub-zero temperatures. Cirrus clouds, in higher latitudes often found at elevations around 3500 m, in the tropics rarely occur below 6000 m. Medium-high clouds, which in cooler climates are limited to elevations below 5000 m, in the tropics often reach levels as high as 7500 m.

There is also a difference between the tropics and the higher latitudes in the frequency of occurrence of the various cloud types. The main variation is caused by the paucity of stratiform clouds in the tropics, while in the mid-latitudes this is the prevailing type in areas near the Polar Fronts. Stratus clouds in the tropics are limited to areas of widespread convergence, in easterly waves and other disturbances, and to low level fog caused by air drainage at night or over cold ocean surfaces.

Figure 7.8 A cumulus cloud with beginning anvil head, caused by the orographic lifting of north-east monsoon air masses over the mountains of central Malaya. (November, 1966)

Figure 7.9 While convectional cumulus clouds form over the land, the lake surface is not sufficiently heated to produce updraughts. (Lake Kariba, Zambia, May 1970)

The dominant cloud type in the tropics is cumulus, in all its various forms. These clouds are always related to vigorous upward movements of air. Frequently the buoyancy related to convectional or conditional instability causes these rising air currents, but cumulus clouds can also be the result of orographic lifting or local convergences, such as along a sea-breeze 'front'. Tropical disturbances are usually accompanied by many cumuli (Figures 6.3 and 6.6) and the trade winds have their own typical cumulus clouds the flat top of which indicates the bottom level of the inversion (Figure 4.6).

The speed of the upward movement, and the vertical distance over which it prevails, control the type of cumulus cloud that is formed. When the rising current has a speed in the order of 2–5 metres per second and is limited in height,

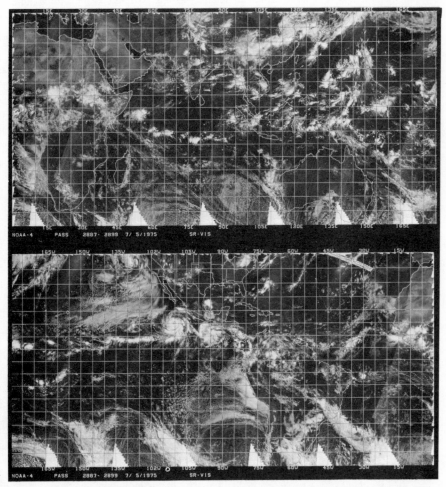

Figure 7.10   Amorphous cloud masses, vortex patterns and cloud bands in the tropics. (N.O.A.A. satellite photograph)

fair weather cumuli (*Cumulus humilis*) are formed. Stronger uplifts, persisting over thicker layers of air, will cause towering cumuli (*Cumulus castellanus*). When the uplift reaches velocities in the order of 20–30 metres per second, cumulonimbus clouds usually result (Figures 7.8, 7.9). This last type, the most impressive of the tropical clouds, can reach levels up to 16 000 m (Riehl, 1954).

Subsiding air movements dissolve clouds, because the air is warmed adiabatically. This can be observed in the subtropical high pressure cells, where clouds are few, and on the leeward side of mountain ranges.

Satellite photographs survey the distribution of clouds over much larger areas than can be observed from the earth's surface (Figure 7.10). The pictures emphasize the horizontal dimensions rather than the vertical thickness of clouds. Three major distribution patterns prevail in the tropics: amorphous cloud masses, showing no distinct structure, vortex patterns and cloud bands. Amorphous cloud masses are often formed in trade wind areas, their basic units are the trade wind cumuli. Vortex patterns are usually associated with cyclonic disturbances (Figure 6.7). Cloud bands are found in zones of convergence or horizontal shear, or near linear systems (Conover *et al.*, 1969).

Cloudiness

The proportion of the sky which is covered by clouds, independent of type, is indicated in eighths or tenths as cloudiness. Its observation is rather subjective and different observers estimate the same amount of cloudiness at different values. Complementary information about cloudiness can be obtained by the observation of actual hours of sunshine, which is much less subject to observer's errors of estimate. The simple formula: $S + C = 100$, in which S is the actual sunshine as a percentage of its maximum possible duration, and C is cloudiness, expressed as a percentage of the visible sky, gives a reliable approximation (Conrad and Pollak, 1962). But all data obtained are for daytime only.

The rarity of stratiform clouds is the main reason for the comparatively low cloudiness figures in the tropics (Figure 7.11). The slightly higher values near the equator are related to the I.T.C.Z. and its associated clouds, the very low mean around 20–30 degrees latitude are caused by the subtropical high pressure cells with their subsiding air movements.

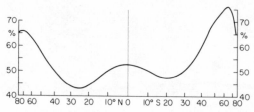

Figure 7.11   Mean annual cloudiness at various latitudes (after Sellers, 1965)

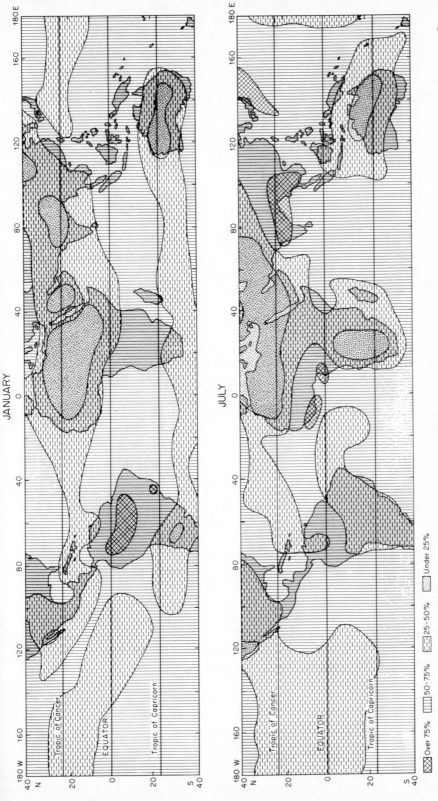

Figure 7.12 Mean daytime cloudiness in January and July (after Landsberg, 1945)

Within the tropics, cloudiness varies strongly, with respect both to location and to seasons (Figure 7.12). In *January* the general pattern of much cloudiness near the I.T.C.Z. and low cloudiness percentages over the subtropical highs prevails (compare Figure 4.2). Deviations from this pattern occur mainly over the continental areas of northern Africa and southern Asia, with rather low cloudiness values, and over the southern Atlantic Ocean, where cloudiness is stronger than expected. In *July* the distribution is quite different. The monsoonal areas of West Africa and southern and eastern Asia show much cloudiness, but the equatorial areas over large parts of the Pacific and Atlantic Ocean, and even over most of the Indonesian archipelago show rather low cloudiness figures. Most of Africa and southwestern Asia also show little cloudiness in July.

It should be pointed out here, that there is little correlation between the distributions of cloudiness and rainfall, as a comparison with rainfall maps (Figure 8.2) will confirm. A striking example is given by the coastal areas of Namibia, Peru and Morocco, which have relatively high cloudiness figures, but very little precipitation, because the persistent stratus clouds are usually too thin to produce rainfall.

The diurnal variations of cloudiness is different over land and over sea surfaces. Over land an afternoon maximum of cloudiness, caused by convectional processes, prevails in most of the tropics. Minimum cloudiness occurs at night, when stability is restored as ground temperatures decrease. However, when stratiform clouds are present in the evening, cooling during the night is retarded and the cloud cover may then persist throughout the night. A secondary maximum of cloudiness, caused by morning fog, occurs in more continental areas, especially over valleys and basins.

Over water surfaces, cloudiness tends to show a maximum at night, when instability is increased by radiation cooling from cloud tops. The minimum of cloudiness is reached during the late morning, when direct absorption of insolation by the lower layers of air warms these layers sufficiently to dissipate low clouds. In the afternoon some convectional clouds may develop even over the sea (Watts, 1955, pp. 39–40).

Because of these differences, coastal areas often show rather complicated diurnal cycles of cloudiness, depending on the form of the coastline, local wind systems, topography and the characteristics of the prevailing air masses. In many places two maxima occur: one in the morning, caused by clouds drifting in from the sea with the beginning sea breeze, and the other one during the afternoon, related to convection and the sea breeze front over land. The minimum is usually at night, when temperatures decrease and the land breeze frequently causes a slight subsiding air movement.

References

Barry, R. G., 1969, The world hydrological cycle, in Chorley, R. J. (Ed.), *Water, Earth and Man*, London, Methuen, pp. 12–13.

Conover, J. H., Lanterman, W. S. and Schaefer, V. J., 1969, Major cloud systems, in Rex, D. F. (Ed.), *Climate of the Free Atmosphere*, (Volume 4 of *World Survey of Climatology*), Amsterdam, New York, Elsevier, pp. 232–243.

Conrad, V., and Pollak, L. W., 1962, *Methods in Climatology*, 2nd ed., Cambridge, Mass., Harvard University Press, pp. 58–60.

Landsberg, H., 1945, Climatology in Berry, F. A., Bollay, E. and Beers, N. R., (Ed.), *Handbook of Meteorology*, New York, McGraw-Hill, pp. 928–997.

Nieuwolt, S., 1966, The urban microclimate of Singapore, *Journal of Tropical Geography*, **22**, 31, 34.

Nieuwolt, S., 1968, Uniformity and variation in an equatorial climate, *Journal of Tropical Geography*, **27**, 37–38.

Penman, H. L., 1948, Natural evaporation from open water, bare soil and grass, *Proc. Royal Society*, Series A, **193**, 120–145.

Penman, H. L., 1949, The dependence of transpiration on weather and soil conditions, *Journal of Soil Science*, **1**, 74–89.

Riehl, H., 1954, *Tropical Meteorology*, New York, McGraw-Hill, pp. 131–134.

Sellers, W. D., 1965, *Physical Climatology*, Chicago, University of Chigaco Press, 272 pp.

Thornthwaite, C. W., 1948, An approach toward a rational classification of climate, *Geographical Review*, **38**, 55.

Tison, L. J., 1960, Aspects hydrométéorologiques des forèts équatoriales et tropicales, in Bargman, D. J. (Ed.), *Tropical Meteorology in Africa*, Nairobi, Munitalp, pp. 327–329.

Watts, I. E. M., 1955, *Equatorial Weather*, London, University of London Press, 224 pp.

Wiesner, C. J., 1970, *Hydrometeorology*, London, Chapman & Hall, p. 88.

Wüst, G., 1922, Verdunstung und Niederschlag auf der Erde, *Zeitschrift der Gesellschaft f. Erdkunde, Berlin*, pp. 35–43.

Wüst, G., 1936, Oberflächensalzgehalt, Verdunstung und Niederschlag auf dem Weltmeere, *Festschrift f. N. Krebs*, Stuttgart, pp. 347–359.

CHAPTER 8

Tropical Precipitation

Atmospheric precipitation in the tropics consists almost entirely of rainfall. It is the most variable element of tropical climates. Almost everywhere in the tropics its most important quantitative indicator, the annual total, differs widely from year to year and in addition it varies strongly with place as well. Other rainfall characteristics, such as its seasonal and diurnal distribution, intensity, duration and frequency of rain-days, also show important differences, in respect both of place and time, in most areas with tropical climates.

Because the other elements of climate are much more uniform in the tropics, rainfall is the main factor used in the delimitation and subdivision of tropical climates (Chapters 1 and 9). Rainfall is also of great practical importance in the tropics, and this is particularly the case in relation to agriculture, as will be explained in Chapter 10.

For these reasons, the need of reliable rainfall data is much larger than for other climatic elements in the tropics. This is widely recognized and in most tropical countries the number of rainfall stations is a few times larger than that of the more complete meteorological stations, where the other data are collected. Nevertheless it is often difficult to obtain a clear picture of rainfall conditions in many parts of the tropics (Riehl, 1954, pp. 72–74). This is especially the case over large parts of the tropical oceans where only a few rainfall recordings are made, mostly on islands, where conditions are not truly representative of large regions. But also over many tropical land areas the network of rainfall stations is not dense enough to produce a satisfactory picture of the variability of rainfall with place.

In addition, in many places rainfall records have been collected over relatively short periods, and they are insufficient to give a reliable indication of the mean or normal values and their variability over time. Moreover, some of these records are of questionable quality, because of the great lack in many tropical countries of well-trained and reasonably paid observers and of proper instruments.

The need for more data about tropical rainfall is particularly acute in some highland areas, where rainfall varies strongly with elevation and exposure, and where climatic favours have attracted a dense population and where agricultural activity is therefore quite intensive.

Because of this shortage of statistical information, the following description of tropical precipitation is based on many interpolations and estimates. While the data for single stations are reliable, the world maps should not be considered accurate in detail, but should rather be taken as indicators of general patterns.

The origin of tropical rainfall

Ultimately, all atmospheric precipitation is the result of upward movements of air, which cause cooling by expansion beyond the level of condensation. This is true in all climates, but there are two major differences in this basic process between tropical and extra-tropical areas.

The first is related to the main cause of the rising air movements. In the mid-latitudes the uplifting is frequently related to cyclonic activities and most of it takes place along frontal surfaces. In the tropics, however, the main cause of rising air is convection, which is responsible, alone or in combination with other factors, for most tropical rainfall, as indicated by the omnipresence of cumulus clouds and thunderstorms. Therefore, upward air movements in the tropics are often of higher speed but of shorter duration than in the mid-latitudes (Malkus, 1964). Tropical air uplifts are also frequently restricted to small areas, because they are limited by the size of the convection cells.

This difference between tropics and mid-latitudes is, of course, not an absolute one. Convectional uplifts also occur in the extra-tropical latitudes, though they are limited to the summer. Slow upward movements of air also take place in the tropics, where they may be caused by large-scale convergence. Orographic uplifts are much the same in all climates. Nevertheless, the difference of degree is a real one and it is of significance in relation to a number of characteristics which distinguish tropical rainfall from extra-tropical precipitation.

The second difference is caused by the properties of the air masses which are uplifted. In the tropics these are generally warmer and more humid than outside the tropics. Warm and humid air masses reach the condensation level at relatively high temperatures and consequently form clouds which consist mainly of liquid water droplets and rarely of ice crystals. Tropical clouds frequently produce copious rainfall while they are at elevations where the temperature is well above the freezing point. This is the 'warm rain', in the formation of which the ice phase of cloud particles is of limited significance. This is in contrast to conditions in the mid-latitudes, where the ice phase causes the growing and coalescence of cloud droplets by the Bergeron effect.

To explain this difference, it is assumed that the rapid condensation which takes place when humid and warm air masses are uplifted at high speeds creates cloud droplets of widely different sizes. A few very large ones whill have started around giant and strongly hygroscopic nuclei. As these cloud droplets of different sizes move at different speeds, they collide frequently and coalescence is the main consequence, resulting in rapid growth to the size of raindrops, often of very large ones (Wexler, 1954).

These two differences in origin explain many features of tropical rainfall: it is generally more intensive, of shorter duration and of stronger localization than outside the tropics. Its diurnal and seasonal distribution and variability are also related to its main processes of origin, though many other factors are at work.

Total rainfall

The total amount of rainfall received at a specific station is normally computed on an annual basis and indicated by the long-term *annual mean*. This is undoubtedly the most widely used rainfall figure. The total amount of rainfall varies, of course, from year to year and the annual mean should therefore be based on long periods of observation to exclude the effects of variability over time. In most parts of the tropics a minimum of about 30 years is required to obtain a reliable indication of the real long-term mean. In some areas, where variability is particularly strong, even longer periods are necessary (Wiesner, 1970). However, at many tropical rainfall stations records have not yet reached sufficiently long periods and the annual mean of some of these stations cannot be considered as much better than an estimate.

The annual mean is generally accepted as the most important indicator of rainfall conditions. Its calculation is simple, which prevents computing errors, and a large public will easily understand its meaning, though not its limitations. For the annual mean has some serious disadvantages when it is used to estimate future rainfall, as for instance in water balance equations and predictions of agricultural possibilities. At most rainfall stations in the tropics, as in other climates, the negative departures from the annual mean are more numerous than the positive ones, resulting in a positive skewness of the frequency distribution of rainfall totals for individual years. The annual mean is therefore inflated by a few very high annual totals and indicates amounts which are larger than can normally be expected with a probability around 50 per cent. The skewness in the frequency distribution is generally strongest where rainfall totals are low, and where a correct estimate of the rainfall that can be expected is most important (Riehl, 1954, pp. 90–91).

For purposes of predicting future rainfall amounts, medians and other data based on probability, such as quartiles, quintiles and deciles, are much better. However, their computation is rather cumbersome and these data are meaningless unless based on long records of at least 20 years. For most stations in the tropics they are not available in published form. Rainfall maps and statistics on a world scale, or for a large number of stations, are therefore almost always based on annual means. A standard period of 30 years is now internationally used (WMO, 1971).

When annual means are compared to latitude, they indicate a clear maximum in the tropics (Figure 8.1). The location of this maximum corresponds almost exactly to the average annual latitudinal position of the I.T.C.Z. over most of the oceans (compare Figure 4.2). This is not surprising, because the I.T.C.Z.,

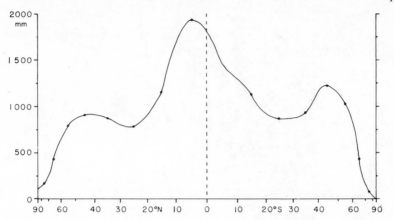

Figure 8.1   Latitudinal variation of mean annual precipitation (after Sellers, 1965)

with its widespread upward movements of warm and humid air masses is the most effective rain-producting part of the general circulation. The minimum of annual rainfall at latitudes between 20 and 30 degrees is caused by the subtropical high pressure cells and the adjacent stable portions of the trade winds, which inhibit rainfall because they have predominantly subsiding air movements. The little rainfall in these latitudes is mainly received in the monsoonal climates. Compared with the polar minima, the low amounts of rainfall in the outer tropics are still quite considerable.

The horizontal axis of the diagram is divided according to the surface of each latitudinal belt of ten degrees. The graph therefore illustrates the large proportion of the total global precipitation that is received in the tropics, which amounts to about 50–60 per cent (Figure 8.1).

The distribution map of annual mean rainfall illustrates these latitudinal zones very well, but it also indicates many non-latitudinal differences within the tropics (Figure 8.2). High amounts of rainfall are caused by a number of factors, usually in combination. The most important of these is, of course, the I.T.C.Z. and the length of its stay over a certain area. This is demonstrated by the high totals over the Pacific Ocean, just north of the equator, where the I.T.C.Z. is nearby almost throughout the year. Over the Atlantic Ocean a similar zone exists, but it is of lower intensity as the I.T.C.Z. is here more variable in position (Figure 4.2).

A second factor causing high rainfall totals is relief. Orographic lifting is particularly efficient where monsoonal winds are forced to rise, as illustrated by the western coasts of India, Burma, Sumatra and Borneo, but also over the Ganges plain of northern India. In West Africa this effect is demonstrated by the coastal areas of Liberia and Sierra Leone, and near Mount Cameroun. The trade winds can also yield large amounts of rainfall when uplifted by steep mountain ranges, as shown along the eastern coasts of Madagascar and the northeastern parts of South America.

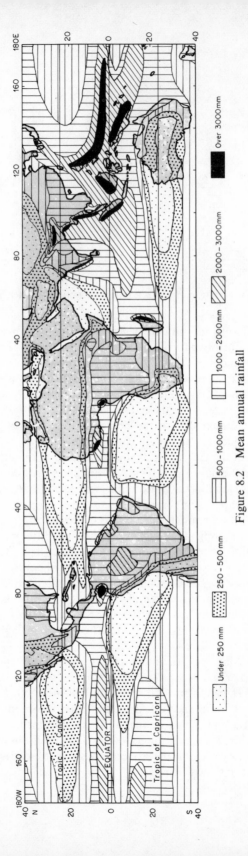

Figure 8.2　Mean annual rainfall

Under 250 mm

250 – 500 mm

500 – 1000 mm

1000 – 2000 mm

2000 – 3000 mm

Over 3000mm

Tropic of Cancer

EQUATOR

Tropic of Capricorn

Where these two factors are combined, very high rainfall results: over New Guinea, near the west coast of Central America and the western parts of the Amazon basin.

A third factor causing high totals of rainfall are the tropical cyclones, in areas where they occur frequently, such as the Caribbean and especially the western edges of the Pacific Ocean, where the zone of high rainfall follows the recurving paths of the storms.

A fourth, and minor, factor is related to the slowing down and changing of direction of the trade winds when they approach the equator. Small areas over the Pacific and Indian Oceans south of the equator are evidence of this factor, they correspond to the average position of the southern branch of the I.T.C.Z..

Areas where rainfall totals are low are mostly caused by the effects of the subtropical high pressure cells, as clearly shown by their location (Figure 8.2). These influences are reinforced by cold ocean currents in the eastern parts of the large ocean basins. Continentality is another factor causing low rainfall, as demonstrated in south-west and central Asia.

The effects of elevation

As shown by the world map (Figure 8.2), mountains and highlands receive more rainfall than nearby lowlands, at least on their windward sides. This is the result of orographic lifting, a process which increases rainfall in all climates. But the effects show a significant difference between the tropics and the extra-tropical latitudes: while outside the tropics the amounts of precipitation increase with elevation up to the highest levels of the mountains, in the tropics the increase stops at a level of about 1000–1500 m, and above this elevation precipitation generally decreases with height. This characteristic is due to two conditions, which frequently prevail in the tropics, but which are the rather rare in the mid-latitudes. The first is a strong difference in water vapour content between the lower and upper layers of the troposphere. Tropical air masses are often very humid up to an elevation of about 800–1500 m, but above this level they are usually rather dry. This may be due to the trade wind inversion, but it can also be caused by the extremely large water vapour production at the earth's surface in the tropics. The steep lapse rate frequently present in tropical air masses also tends to reduce the capacity of the air to retain water vapour in the higher parts of the troposphere.

The second factor is the predominance of vertical air movements in the tropics, where horizontal advection of moisture is often limited. Most precipitation therefore originates from the atmosphere directly above the slopes, while in the mid-latitudes water vapour is often transported over large horizontal distances (Weischet, 1965, 1969).

These factors explain why higher parts and interior regions of extensive highlands, over which the lower, humid parts of the troposphere are absent, generally receive much less precipitation than the lower and outer slopes.

The variability of annual rainfall over time

The variation of the annual rainfall from year to year is a very important feature of climate, which is unfortunately often disregarded, namely when the annual

108

mean is simply accepted as the sole indicator of rainfall. At most tropical stations the annual total varies widely (Figure 8.3). The illustrated stations, which show variations between 50 per cent and 150 per cent of the annual mean over a period of 30 years, are quite representative of tropical conditions. The diagram shows that negative departures from the mean are more numerous than the positive ones. The three stations all have a ratio of 16 to 14, but in the drier parts of the tropics this ratio would be much higher, as the skewness in the frequency distribution is more pronounced.

The graph also indicates that dry years tend to have a certain amount of persistence. Annual totals below the long-term mean come frequently in series of three, four or even five dry years (Nagpur, 1950–1954). This is a feature common to many parts of the tropics.

This variability over time is the main reason why the annual mean should be based on long periods of records. But some information about its variability should be given wherever possible, as the annual mean alone provides only a rather crude and unreliable indication of rainfall conditions.

The simplest indicators are the maximum and minimum annual totals, but the significance of these data depends entirely on the length of records and they do not allow comparisons between stations unless the periods of observation are the same. A better indicator is the mean variability, computed as the average departure from the mean, disregarding whether these are positive or negative. Another useful figure is the standard deviation of the annual mean. Where the annual totals have a very skew frequency distribution, the quartile deviation or similar indicators should be used.

These absolute figures of variability can only be used when a few stations in the same climatic area are compared, or to illustrate conditions at one single

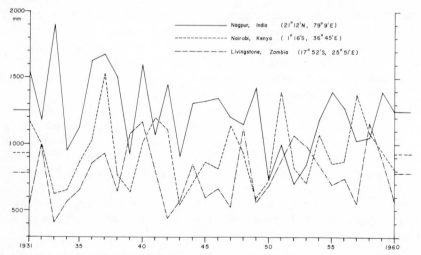

Figure 8.3 Annual rainfall totals for the period 1931–1960 at three tropical stations (means for the same period are indicated in the margins)

station. But when regional differences are studied, variability must be indicated in relative terms. A mean variability of 100 mm is a small one when the annual mean is around 1000 mm, but it is rather significant when the annual mean reaches only 300 mm. For the purpose of regional comparisons, and for the presentation of variability of maps, it should be indicated in relative terms, usually as a percentage of the annual mean or median. Most widely used for these purposes is the 'coefficient of variability', which expresses the standard deviation in this way. The mean relative variability gives the mean departure from the annual mean as a percentage of that mean.

The mean relative variability of annual rainfall on a world scale is often illustrated by a frequently reproduced map (Biel, 1929). On this, and similar maps, the humid tropics stand out as regions with a very low variability of rainfall. However, this picture is somewhat misleading, because the units of measurement, 1 per cent of the annual mean, are much larger in the humid areas than in dry zones.

From a practical point of view, rainfall variability is of little consequence in very humid areas, where there is usually enough water available for all purposes, including agriculture and hydroelectric use. Even in the humid parts of the tropics relatively dry years can occasionally occur, and these can be indicated by variability figures, but they cannot be predicted on a statistical basis and must be considered part of the normal climatic risk. Variability of annual rainfall is also of little significance in the dry climates, where both domestic and agriculturalral water supply are based on other sources of water, rather than on rainfall.

The variability of annual rainfall is, however, of crucial importance in those parts of the tropics where rainfall is marginal or barely sufficient for normal agricultural use. Here a sequence of dry years may mean death and starvation for most inhabitants and their animals, as was recently dramatically shown in the Sahel region of West Africa and other areas. In these areas the effectiveness of the little rainfall that comes depends also on its intensity and seasonal distribution.

The seasonal distribution of rainfall

In the tropics the seasonal regime is second in importance only to the total amount of rainfall. It is the major controlling factor of the calendar of agricultural activities in most tropical climates (Chapter 10). In many parts of the tropics the times of the start, duration and end of the rainy season are decisive in the struggle for sufficient food supply. Rainy seasons also bring different temperature, moisture and cloud conditions compared to the dry periods, and they therefore influence the general weather conditions. They even have a strong effect on the way of life, because outdoor activities are much more prevalent in the tropics than in most other climates and they are often hampered by rain.

110

Because other climatic elements are much more uniform, the seasonal rainfall distribution forms the basis of many classifications and subdivisions of the tropical climates (Creutzburg, 1950; Koeppen, 1936; de Martonne, 1941; Thornthwaite, 1948; Troll, 1964; von Wiszmann, 1948).

Rainfall regimes in the tropics are controlled by a number of factors, some of which are active over very large areas, while others are effective over much smaller regions. The first group of factors consists of elements of the general circulation of the tropical atmosphere. The main one is, of course, the I.T.C.Z., in the vicinity of which much rainfall is always received (Riehl, 1954, pp. 76–79). The subtropical high pressure cells constitute the second major inhibiting factor on rainfall. Their influences are carried towards the equator by the trade winds.

It is possible to develop a simple model of rainfall regimes in the tropics based on the effects of these two major factors (Espenshade, 1973; Petterssen, 1969). This model consists of three latitudinal belts: a zone of continuous high rainfall

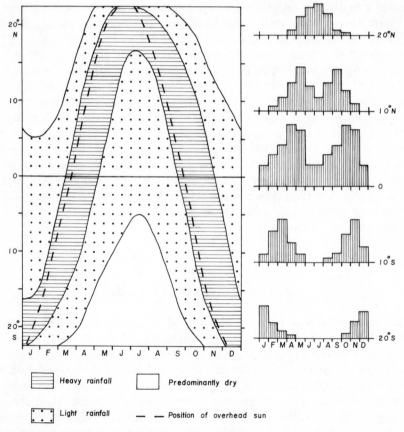

Figure 8.4 A simplified model of seasonal rainfall distribution over tropical continents. Left: latitudinal movement of precipitation zones Right: station models at various latitudes (partly after Miller, 1971)

near the equator, where the I.T.C.Z. is close by throughout the year, a zone of very low rainfall around latitudes from about 15 to 25 degrees, where the subtropical high pressure cells are situated, and an intermediate belt, where the effects of these two factors alternate seasonally.

Over the oceans, where both the I.T.C.Z. and the subtropical high pressure cells change their positions only over small distances in the course of the year, the intermediate belt is rather narrow. In continental areas, however, where the seasonal displacements of the I.T.C.Z. are large, the theoretical pattern of the seasonal rainfall regimes would be more complicated (Figure 8.4). According to this model, rainfall near the equator would be continuous throughout the year, but with two maxima and two periods of less rain. The maxima would occur approximately one month after the equinoxes, when the I.T.C.Z. is directly over the equator. As these periods of heavy rainfall seem to follow the seasonal displacement of the overhead position of the sun, they are sometimes called 'zenithal rains'. It should be realized, however, that it is the convergence of air masses which produce the rains, and not the increase in intensity of the solar radiation, which is too small to affect the main rain-producing processes (Figure 2.4). Therefore the term 'zenithal rains' is better avoided.

With increasing distance from the equator, the two seasons of maximum rainfall move together in time, reducing one minimum in length, while the other dry season becomes longer and intenser. The long dry period always comes during the 'winter' half year, when the sun is overhead over the opposite hemisphere (Figure 8.4, station models for 10° N and S). At the same time the total annual amount of rainfall is less than near the equator. At a latitude of about 15 degrees the short dry season disappears completely and the two maxima have merged into one.

Further away from the equator the period of maximum rainfall becomes weaker and shorter, as the effects of the I.T.C.Z. are less noticeable and the influence of the subtropical high pressure cells predominates during most of the year (Figure 8.4, station models for 20° N and S).

Over oceanic areas the actual pattern of the seasonal rainfall distribution corresponds fairly well to this model. Only two minor factors sometimes produce irregularities. The first, tropical cyclones and easterly waves, bring rainfall in trade wind areas which are normally rather dry. The second, remnants of extra-tropical disturbances which occasionally travel to the low latitudes, especially over warm ocean currents, bring irregular rainfall. However, these two minor factors affect only relatively small areas and then for short periods.

Over the tropical continents the actual variations of the seasonal rainfall distribution are much more complicated than suggested by the model (Figure 8.5). This is caused by a number of factors which locally or regionally interfere with the general pattern. The most effective of these is convection, which can increase the amounts of rainfall locally. In the outer tropics convection is more prevalent during the period of the overhead sun, but nearer to the equator it occurs throughout the year. A second factor is orographic lifting, which can produce enormous amounts of rainfall in the tropics. On the

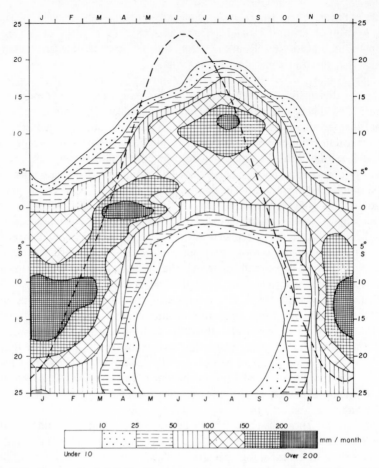

Figure 8.5 Seasonal rainfall distribution over eastern Africa, along 32° E, as indicated by monthly means. The dashed line is the path of the overhead sun (after Flohn, 1964)

other hand, rain shadow effects on the leeward sides of mountain ranges are also pronounced. Where the predominant wind directions change seasonally, as in the monsoonal climates, this factor creates large regional differences in rainfall regime. These differences are often intensified by the characteristics of the air masses which are uplifted, as for instance in India and China, where the two monsoon winds bring entirely different air masses, which as a consequence produce very different amounts of rainfall.

The combined effects of these major factors and the regional influences produce a complicated pattern of rainfall regimes over the tropical continents. There are numerous variations, but it is possible to generalize these to three major types of region:

(1) The continuously rainy areas, where there is some seasonal variation in

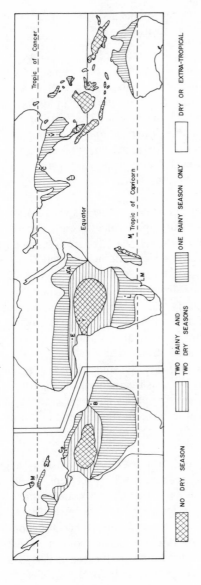

Figure 8.6 Seasonal rainfall distribution types in the humid tropics. Letters indicate stations for which rainfall diagrams are shown in Figure 8.7 (partly after Blüthgen, 1966)

rainfall, but where no real dry season occurs. This category prevails where the mean rainfall during the driest month of the year exceeds 60 mm.

(2) Areas where two rainy seasons and two dry periods alternate. The intensity and duration of the two rainy periods usually differ, and the two dry seasons are also rarely of the same length.

(3) Areas where only one rainy season occurs. Almost everywhere it comes when the sun is close to its overhead position.

The areal distribution of these three types shows some conformity to the theoretical model based on the two major factors (Figure 8.6). The continuously rainy type is found mainly in the equatorial areas, though it also prevails in the eastern parts of Madagascar, where orographic uplifting of trade wind air masses continues almost throughout the year. The second type is confined to the continents of Africa and South America, and it generally occurs at slightly higher latitudes, surrounding the first type. The type with only one rainy season generally prevails at the higher latitudes and in areas with monsoons. It occurs, however, very close to the equator in the eastern parts of Africa and South America.

The types of rainfall regime indicated on the map were deliberately described in very general terms, to avoid an impression of accuracy which is unwarranted, not only because insufficient data are available from many parts of the humid tropics, but also because the boundaries fluctuate widely from year to year. The use of only three categories also means that each type covers a very large area, which contains many regional variations. To illustrate the extent of these differences within each major type, a number of station diagrams are provided (Figure 8.7). On these diagrams the monthly means are all expressed as a percent of the annual mean for the same station. The stations are grouped according to their latitudinal position.

The first group of stations are located at latitudes below 5 degrees (Figure 8.7(a)). Only Yaounde shows the theoretical equatorial regime with two maxima and two minima of approximately equal intensity and duration. Belem, though very close to the equator, has only one dry and one rainy season. Penang displays two of each, but their intensities are different, mainly because of monsoonal influences.

The second group of stations, situated between 5 and 10 degrees of latitude, also illustrates large differences (Figure 8.7(b)). Here, Enugu confirms fairly well to the theoretical type as suggested by the model (Figure 8.4). At Caracas the rainy season shows almost no interruption. In Addis Ababa the concentration of rainfall is very strong: during the four months from June to September almost three-quarters of the annual total is normally received.

At the third group of stations, which are located at latitudes between 15 and 20 degrees, one rainy season dominates (Figure 8.7(c)). Only Vientiane shows a slight interruption. Lusaka, very continental, has an absolutely dry season which lasts for 4 months and over 90 per cent of the annual total rainfall comes during the period from November to March. Mauritius, in a rather oceanic position, has a much wetter period of minimum rainfall.

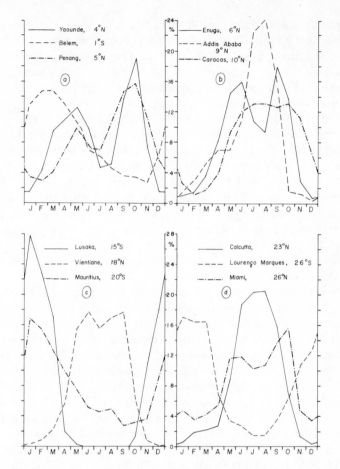

Figure 8.7   Seasonal rainfall distribution at various tropical stations (monthly means as a percentage of the annual mean). The location of these stations is indicated on Figure 8.6

The fourth group of stations, between 23 and 26 degrees, also shows only one rainy and one dry season (Figure 8.7(d)). The strong maximum in Miami, Florida, during September and October is caused by hurricane activity, which reaches its maximum during this period. The strong summer concentration of rainfall in Calcutta is, of course, the result of the summer monsoon.

Regional variations of the seasonal rainfall distribution are described in many textbooks of meteorology and climatology (Boucher, 1975; Riehl, 1954, pp. 79–87). They will be used in the next chapter of this book as an important feature for the subdivision of the tropics (Chapter 9).

The use of monthly means to indicate the seasonal rainfall distribution brings a number of disadvantages. First, monthly means, like the annual ones, tend to be inflated by a few heavy rainfalls, and they are therefore often not very representative of normal conditions. This is particularly the case for relatively dry months (Riehl, 1954, pp. 90–91). Secondly, monthly means give no indication of the rainfall variability from year to year, which is rather strong where short periods, like a month, are used. Thirdly, events which do not occur every year during the same month are not shown by monthly means, as they are averaged out. Events which last considerably shorter than one month are, of course, also not represented.

Most of these disadvantages can be avoided by using monthly medians, quartiles or other indicators based on order of magnitude rather than averages. The seasonal rainfall distribution can be expressed in terms of reliability and probability and its description can be made more truly representative than when monthly means are used (Evans, 1955; Glover and Robinson, 1953, 1954; Kenworthy and Glover, 1958; Nieuwolt, 1973a). However, these data are rarely published and their computation is rather laborious because long periods of observation must be used. Where large numbers of stations are compared, or where stations from different countries must be used, this is obviously impracticable. In these cases the researcher has to refer to monthly means, which are directly available for most stations.

Events which do not occur regularly every year at about the same time can only be traced by checking the data for individual years. In this way the interruption of the rainy season in Tanzania, which was insufficiently indicated by the monthly means at many stations, could be represented by the percentage of all years which showed this interruption (Nieuwolt, 1974a, pp. 191–192).

To represent events which last considerably shorter than one month, shorter units, like 10-day or 5-day periods, must be used. However, this would increase the amount of computation, with greater risk of errors. It would also create some statistical background noise as even during rainy seasons many of these periods would record no rain. Publication of data for these shorter periods would need introduction to a public that is accustomed to monthly figures. It seems that the improved representation of short events is not worth the amount of extra work involved, at least for most purposes (Crowe, 1971, pp. 129–130).

Finally, the use of monthly means or other monthly figures creates difficulties when the regional pattern of the seasonal rainfall distribution is to be shown on a map. As the various regimes are indicated by twelve figures, an almost endless number of variations is possible. One solution is to group similar variations into a few categories, which generalize the regional types (Figure 8.6). Another method is to devise meaningful indices, which may be based on monthly means. The areal distribution of such indices can be represented cartographically by a system of isolines (Nieuwolt, 1974a, pp. 190–191).

The diurnal variation of rainfall

While its significance is definitely less than that of the seasonal distribution, the diurnal rainfall regime is nevertheless an important feature of tropical climates. It has a strong effect on communications, especially where roads are unpaved, and on air traffic; it controls general weather conditions and many outdoor activities, and it influences the efficiency of rain in agriculture, because rain falling during the hot hours of the day is subjected to heavy losses by evaporation.

Still, the number of publications about the diurnal rainfall distribution is rather small. One reason for this is the limited amount of interest in this characteristic in the mid-latitudes, where it is a rather unpredictable factor, because disturbances and fronts cause a strong irregular tendency. However, at most tropical stations the diurnal variation of rainfall is much more regular. Apart from the infrequency of fronts, the prevalence of diurnal processes in the tropical trophosphere is the reason for this difference (Steinhauser, 1966). In the tropics, predictions of rainfall can often by guided by the diurnal regime.

But even in the tropics few studies of the diurnal rainfall variation exist, because suitable data are lacking. It can only be indicated by hourly or, at least, three-hourly, figures which are obtained from automatic raingauges and the number of these instruments is still rather limited in most tropical countries. Moreover, many types of automatic raingauges do not perform very well in the tropics, either because the high humidity disturbs the clockwork mechanism, or because of interference by insects, fungi or larger animals, theft or damage by vandalism, poor technical maintenance and lack of spare parts. Therefore, many hourly rainfall figures are of doubtful reliability. One way to reduce the effects of this factor is by using long periods of observation, but to obtain these will take a lot more time as many rainfall stations in the tropics have only recently been installed.

Two main types of diurnal rainfall regime can be recognized in the tropics:

(1) The *Inland type* displays a maximum of rainfall during the late morning or afternoon. The main reason for this distribution is, of course, convection, caused by surface heating of the land.

(2) The *marine or coastal type* shows a maximum of rainfall during the night or early hours of the morning. This regime is caused by nighttime 'convection', the result of a steepened lapse rate as the upper troposphere is cooled by radiation losses, mainly from the tops of the clouds, while the lower layers of the atmosphere remain warm by close contact with the water surface. During the day, direct absorption of solar radiation heats up the lower layers of the troposphere faster than the water surface, creating stability at low levels and a minimum of rainfall. Other factors which are mentioned to explain this regime are the temperature flow from the water surface to the air, which reaches a maximum in the early parts of the night and again near dawn, and atmospheric tides (Kraus, 1963; Malkus, 1964; Finkelstein, 1964). Some doubts have been cast on the validity of the above explanations, but no generally accepted theory has been substituted (Andre and Bleeker, 1951; Ramage, 1952).

The two types of regime, however, are rarely realized in pure form and only a few stations follow the theoretical pattern described above. Local factors, which have a strong effect on the small-scale processes involved in diurnal rainfall

118

distribution, are the main reason for the many deviations from the ideal type (Thompson, 1965, pp. 1–11).

At inland stations topography is the main cause of local variations in the diurnal regime. Convectional showers develop much earlier over some areas than over others, depending on slope and exposure to the main or local winds. Characteristics of the land, such as vegetation type, drainage conditions, soil humidity and even surface colour, can also cause many variations. Nairobi is a good example of a typical tropical inland station (Figure 8.8). It shows a retarded maximum of rainfall, because storms often develop over the highlands to the north of the city and then drift southwards. Other stations in the same area show much clearer rainfall maxima during the afternoon (Thompson, 1965, pp. 3,4). In Malaya, the inland stations display much stronger afternoon

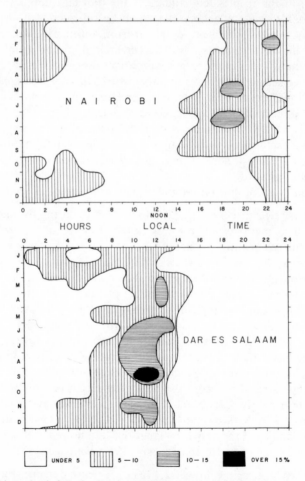

Figure 8.8   Diurnal rainfall distribution at Nairobi and Dar es Salaam (hourly rainfall as a percentage of the monthly mean; three-hour running means). (Source: Thompson, 1965, pp. 4,14)

maxima, but there a system of converging sea breezes over the Malayan peninsula intensifies the normal development of convectional storms over land (Nieuwolt, 1968, pp. 320–326; Ramage, 1964, pp. 66–67). Such converging systems have been observed over many islands and peninsulas, even when as big as Florida (Byers and Rodebush, 1948). Despite all variations, most inland stations show a clear rainfall minimum during the morning and a maximum later in the day. Seasonal differences are relatively small.

At coastal stations the deviations from the ideal marine type are larger. First of all, coastal stations show the marine type only when the prevailing winds of the general circulation are from the sea, and they experience a more inland type of diurnal rainfall regime when the winds come from the landward side. This is the main reason for the large seasonal variations in the diurnal rainfall regime at coastal stations in monsoonal climates (Nieuwolt, 1968, pp. 320–324; Ramage, 1952; Thompson, 1965; Watts, 1955a).

Secondly, many local factors influence the diurnal rainfall distribution. The form of the coastline, the topography of the coastal areas, the presence of swamps, lakes, rivers or irrigated fields all affect the time of rainfall. Their influence is either direct, because they control local convection, or indirect, as they have an effect on the development of the sea and land breezes. The sea breeze itself usually brings little precipitation, because it starts with a stabilizing downward movement of air over the sea. But where it meets other air currents, a 'front' may develop and this is normally an area of thunderstorm activity (Ramage, 1964, pp. 65–66).

The land breeze may also cause rainfall, when it undercuts humid air masses over the sea. And over narrow seas, such as the Straits of Malacca, systems of converging land breezes may develop and cause a nighttime rainfall maximum (Watts, 1954).

Dar es Salaam provides an illustration of the coastal type of regime. It displays local as well as seasonal variations (Figure 8.8). There are many stations along the East African coast which show the coastal type even more clearly (Thompson, 1965, pp. 4,8).

Finally it should be mentioned that the coastal type of diurnal rainfall distribution is also found near great lakes, as for instance near Lake Victoria (Thompson, 1965, pp. 1,8). At these locations topography plays a very important role and it is rare for the coastal type to be clearly developed.

Rainfall frequency

The frequency of occurrence of rainfall is usually indicated by the number of *rain-days*, which are defined as periods of 24 hours with more than a certain amount of rain. This method has many advantages, particularly in the tropics: ordinary raingauges can be used, and they need to be read only once a day; even untrained observers can record a rain-day; the danger of recording errors is small and the computation of totals is very simple.

However, the definition of what constitutes a rain-day is far from uniform: most countries of the British Commonwealth use a lower limit of 0·25 mm (0·01 inch), but other countries have fixed it at 1, 2 or 5 mm. Comparison of rain-day numbers from different countries is therefore in many cases not possible.

Another disadvantage of the method is the lower limit of only 0·25 mm, used in many parts of the tropics. In warm climates, total rainfalls of less than about 2 mm are of almost no significance for agriculture or water supply usage since most of these small amounts will evaporate before infiltrating the soil. Since days with such little rain constitute a large proportion of the total number of rain-days, the value of these figures for many practical usages is seriously reduced.

While it is always possible to use higher limits and produce more meaningful rain-day numbers, in most countries this involves laborious computation from the original daily rainfall figures (Nieuwolt, 1974a, pp. 192–193).

Finally, the use of a period of 24 hours can be criticized because it bears no relation to the actual duration of the rain-producing processes (Riehl, 1954, p. 94). While this is a valid objection, the 24-hour period constitutes a very practical interval, because it establishes a simple routine of reading the raingauge every day at the same hour. It has also been shown that this period represents the cumulative effects of several rainstorms during the same day very well (Nieuwolt, 1974b, pp. 243–245).

A strong correlation exists between the number of rain-days and the total amount of rainfall, though in the tropics the seasonal variations in this correlation can be quite large (Jackson, 1972). Regional differences are usually smaller, especially during rainy months (Dale, 1960, p. 21; Wycherley, 1967).

However, the tropical highlands are an important exception to this general rule; there the number of rain-days increases with elevation even at levels above 1000–1500 m where the total amount of rainfall generally decreases. With the higher number of rain-days, the highlands also show a more regular frequency of rainfall, indicated by the lower variability from year to year, compared to the lowlands (Dale, 1960, pp. 11–12; de Boer, 1950; Nieuwolt, 1973b; Woo Kam Seng, 1968).

The *persistence* of rainfall or of dry conditions can be indicated by *wet or dry spells*, series of consecutive days with or without rain. These spells can be shown in tabulated form, but a comparison between different stations is not easy (Dale, 1960, pp. 23–28). A better method is to illustrate these spells on diagrams (Figure 8.9). They show that dry spells of various lengths are surprisingly frequent in many parts of the tropics, even during rainy seasons. For instance, over a period of ten years, Singapore experienced dry spells of at least seven days during every month of the year except November (Nieuwolt, 1964). These diagrams also indicate the large variability from year to year in rainfall or drought persistence.

The computation of dry spells is particularly useful when the risks of drought in relation to agriculture are estimated (Hutchinson, 1974; Torrance, 1959).

The variability of rainfall from day to day can be expressed as the Interdiurnal Variability Index (Nieuwolt, 1974c). Unfortunately, this index shows no close correlation with the variability of either monthly or annual rainfall.

Rainfall intensity

Intensity relates the total amount of rainfall to its duration. This characteristic is very important in applied climatology, particularly in the tropics where intensities of rainfall are generally high. Rainfall intensity controls the probability and seriousness of local floods, and is therefore a major factor to be

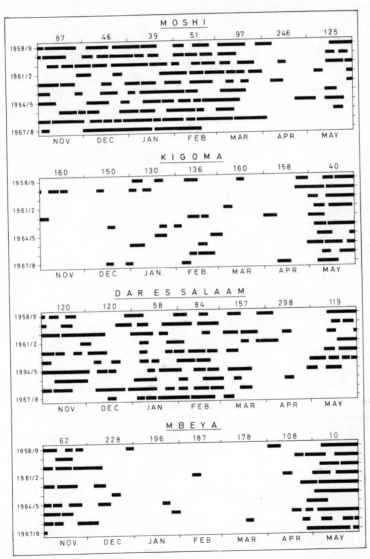

Figure 8.9   Dry spells longer than four days (each day with less than 0·25 mm of rain) at four stations in Tanzania. (Figures on top of each monthly column indicate mean rainfall in millimetres for the illustrated period of 1958–1968.) From Nieuwolt (1974b), with permission from the American Meteorological Society

Figure 8.10   The high intensity of tropical rainfall often produces local floods: an overflowing monsoon drain in Singapore

considered in the planning and construction of dams, reservoirs, drainage canals, culverts and bridges. Moreover, it has a strong influence on the effectiveness of rainfall for agriculture, because when the intensity exceeds the maximum infiltration rate of the soil, surface runoff results and a proportion of the rainfall is lost. Because of this process, rainfall intensity also affects soil erosion, landslides and sedimentation rates in lakes and reservoirs (Rapp, Berry and Temple, 1973).

The *mean intensity* is usually indicated by the *rainfall per rain-day*, using the formula $I = P/N$, where P is the total amount of rainfall and N the number of rain-days. This index can be computed for the whole year or for single months. An idea of annual values both in the tropics and outside is provided by Table 8.1. The table illustrates the high intensity values which prevail in the tropics (Crowe, 1971, p. 140).

Table 8.1   Annual mean rainfall per rain-day, in mm (rain-days with over 1 mm of rain)

Quito, Ecuador	8·5	Accra, Ghana	13·6	Calcutta, India	15·5
Georgetown, Guyana	13·3	Lagos, Nigeria	14·4	Rangoon, Burma	20·9
S. Salvador, El Salvador	16·1	Entebbe, Uganda	12·4	Djakarta, Indonesia	13·5
S. Juan, Puerto Rico	10·1	Bombay, India	22·4	Hong Kong	21·2
		London, England	5·5	Vienna, Austria	4·1

Source: WMO, 1971.

Figure 8.11   Soil erosion in the tropics is often so rapid that the vegetation cover, once destroyed, has no opportunity to re-establish itself. A road cutting near Singapore

The mean rainfall intensity in the tropics shows large seasonal and regional variations (Dale, 1960, pp. 11,12; Jackson, 1972). Generally, intensity increases with the total amount of rainfall and the only exceptions to this rule are found in the highlands at levels over about 1500 m, where the number of rain-days grows with higher elevations, but the total amount of rainfall diminishes, so that the rainfall per rain-day decreases.

The mean rainfall intensity can also be computed for shorter periods than 24 hours, but this necessitates the use of automatic raingauges and is therefore limited to a relatively small number of stations in the tropics. Variations within the same climatic area are quite large when short periods of rainfall are considered, and it seems that orographic lifting is the main factor responsible for these differences (Taylor, 1968; Sellick, 1953).

For most technical purposes, the extremes of rainfall intensity are more significant than its mean. The most commonly used indicator is the *annual series*, in which the extreme values for each year of observation is used. When the maximum rainfall during 24 hours is indicated, this method has the advantage that records are available for many stations in the tropics. The predictions of extremes are usually given in the form of the 'return period' or 'recurrence interval', which is the average interval within which a specified amount of rainfall per 24 hours can be expected to occur once. When the actual values recorded are plotted against their computed return periods on semi-logarithmic paper, they usually form a straight line, so that extrapolation over

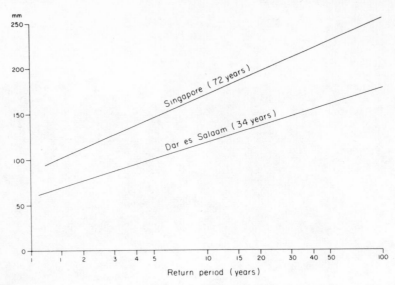

Figure 8.12   Annual series of 24-hours' rainfall at Singapore and Dar es Salaam

longer periods is possible (Figure 8.12). Values at tropical stations are generally much higher than for places outside the tropics (Gilman, 1964).

The same procedure can be used for rainfall intensities during shorter periods, using autographic records (Taylor, 1968; Taylor and Lawes, 1971). The values for different durations at the same station generally show a close correlation (Figure 8.13).

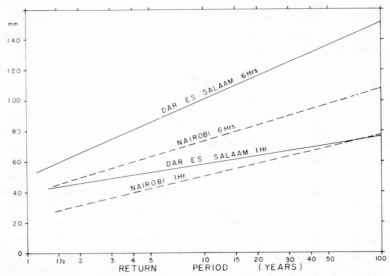

Figure 8.13   Extreme rainfall during 1 and 6 hours at Nairobi (based on 27 years of records) and Dar es Salaam (21 years). (Source: Taylor and Lawes, 1971, p. 13,19)

Figure 8.14   A typical rainstorm over the Great African Rift Valley near Nairobi.
(April, 1974)

Maximum intensities can also be predicted by computing the Probable
Maximum Precipitation, for which various methods have been devised (Lumb,
1971; McCallum, 1965; Sansom, 1965). Values obtained in the tropics are very
high, compared to the mid-latitudes, with the highest values over mountain
slopes rather than interior plateaux (Lockwood, 1967).

Rainstorms

An interesting feature of the tropical climate is the predominance of rainstorms,
defined as relatively short periods of uninterrupted and intense rainfall. Though
this definition gives no limits of duration and intensity, there is usually no
difficulty in identifying a rainstorm. Their duration is rarely more than about
two hours, though some may last much longer when continuous orographic
lifting or a tropical cyclone cause long-lasting heavy rains. Most rainstorms are
approximately circular in shape and their diameter is about 10–16 km, but some
of the more durable ones may be larger in size. Once they are active, rainstorms
usually move only slowly and over short distances, and these movements are
often erratic, though sometimes related to surface features (Orchard and
Sumner, 1970; 1972; Watts, 1955b).
   Rainstorms are important because they contribute up to 90 per cent of the
total rainfall in many parts of the tropics. They also cause a large proportion of
the rain to be concentrated during relatively few days: in most tropical climates

half of the total annual rainfall is received during only 10–20 per cent of the rain-days (Channa, 1968; Riehl, 1954, pp. 95–97). In Dar es Salaam, about two thirds of the rainfall originates from storms which last less than 15 minutes (Nieuwolt, 1974b, p. 242).

Rainstorms are the main reason why the intensity, when measured over the periods of actual rainfall, is much higher in the tropics than in extra-tropical climates (Braak, 1950). They also cause the strong interdiurnal rainfall variability as well as the strong variability of rainfall from year to year, because their erratic movements may bring many more rainstorms to a station during one year than in another. Rainstorms produce a spotty rainfall distribution pattern, not only for any given day, but even over periods as long as one month (Johnson, 1962; Nieuwolt, 1971). When long-term means are mapped, this spottiness is obscured because rainstorms frequently occur in a random pattern.

Rainstorms are the result of convection and other rain-producing processes, which are most effective when they take place in potentially unstable, humid and warm air masses. Such air masses frequently prevail in the tropics and this is the main reason why rainstorms are a typical feature of the tropical climates, whereas they are largely limited to the warm season in the extra-tropical latitudes.

There are relatively few studies of tropical rainstorms, because data on them can only be obtained when a large number of autographic raingauges are concentrated over a small area (Orchard and Sumner, 1970, 1972; Watts, 1955b). Their seasonal and regional variations can, however, be deduced from 24-hourly rainfall records (Nieuwolt, 1974b).

References

Andre, M. J. and Bleeker, W., 1951, On the diurnal variation of precipitation, particularly over the USA, and its relation to large-scale orographic circulation systems, *Quarterly Journal Royal Met. Society*, **77**, 260.

Biel, E., 1929, Die Veränderlichkeit der Jahressumme des Niederschlages auf der Erde, *Geogr. Jahresbericht aus Oesterreich*, **14/15**, 151–180.

Blüthgen, J., 1966, *Allgemeine Klimageographie*, 2nd ed., Berlin, Walter de Gruyter, p. 221.

de Boer, H. J., 1950, On the relation between rainfall and altitude in Java, Indonesia, *Chronica Naturae*, **106**, 424–427.

Boucher, K., 1975, *Global Climate*, London, English Universities Press, p. 78–94, 101–106, 111–123.

Braak, C., 1950, Kenmerkende eigenschappen van het tropische klimaat, *Tijdschrift Kon. Ned. Aardr. Genootschap*, **67**, 623.

Byers, H. R. and Rodebush, H. R., 1948, Causes of thunderstorms of the Florida peninsula, *Journal of Meteorology*, **5**, 275–280.

Channa, J. A., 1968, *An analysis of thunderstorms occurring in Namulonge, Uganda*, paper presented at 4th Specialist Committee on Applied Meteorology, Nairobi.

Creutzburg, N., 1950, Klima, Klimatypen und Klimakarten, *Petermanns Geogr. Mitteilungen*, **94**, 57–69.

Crowe, P. R., 1971, *Concepts in Climatology*, London, Longman, 589 pp.

Dale, W. L., 1960, The rainfall of Malaya, Part II, *Journal of Tropical Geography*, **14**, 11–28.

Espenshade, E. B., 1973, *Goode's World Atlas*, 13th ed., Chicago, Rand McNally, p. 19.

Evans, A. C., 1955, A study of crop production in relation to rainfall reliability, *East African Agricultural Journal*, **20**, 263–267.

Finkelstein, J., 1964, Diurnal variation of rainfall amount on tropical Pacific Islands, *Proc. of the Symposium on Tropical Meteorology, Rotorua, N.Z., 1963*, pp. 286–294.

Flohn, H., 1964, *Uber die Ursachen der Aridität Nordost-Afrikas*, Würzburg, Geogr. Inst., Vol. 12, pp. 25–41.

Gilman, Ch.S., 1964, *Rainfall*, Section 9 in Ven Te Chow (Ed.), *Handbook of Applied Hydrology*, New York, McGraw-Hill, p. 9–55–56.

Glover, J. and Robinson, P., 1953, A simple method of calculating the reliability of rainfall, *E. African Agric. Journal*, **19**, 11–13.

Glover, J. and Robinson, P., 1954, The reliability of rainfall within the growing season, *East African Agricultural Journal*, **19**, 137–139.

Hutchinson, P., 1974, *The Climate of Zambia*, Occasional Study No. 7 of the Zambia Geographical Association, Lusaka, pp. 48–50.

Jackson, I. J., 1972, Mean daily rainfall intensity and number of rain days over Tanzania, *Geografiska Annaler*, Series A, **54**, 369–373.

Johnson, D. H., 1962, Rain in East Africa, *Quarterly Journal of the Royal Met. Society*, **88**, 6–16.

Kenworthy, J. M. and Glover, J., 1958, The reliability of the main rains in Kenya, *East African Agricultural Journal*, **23**, 267–272.

Koeppen, W., 1936, *Das geographische System der Klimate*, Berlin, Gebr. Bornträger, Vol. 1. Part C of Koeppen-Geiger, *Handbuch der Klimatologie*, 44 pp.

Kraus, E. B., 1963. The diurnal precipitation change over the sea, *Journal of the Atmospheric Sciences*, **20**, 551–556.

Lockwood, J. G., 1967, Probable maximum 24-hour precipitation over Malaya by statistical methods, *Meteorological Magazine*, **96**, 18.

Lumb, F. E., 1971, *Probable maximum precipitation in East Africa for durations up to 24 hours*, E.A.M.D., Techn. Memo. No. 16, Nairobi, 8 pp.

McCallum, D., 1965, The relationship between maximum rainfall intensity and time, *E.A.M.D., Memoirs*, Vol. III, No. 7, Nairobi, 8 pp.

Malkus, J. S., 1964, Tropical convection: Progress and outlook, *Proceedings of the Symposium on Tropical Meteorology, Rotorua, N.Z., 1963*, pp. 247–277.

de Martonne, E., 1941, Nouvelle carte mondiale de l'indice d'aridité, *Météorologie*, **37**, 3–26.

Miller, A. A., 1971, *Climatology*, 9th ed., Methuen, London, p. 105.

Nieuwolt, S., 1968, Diurnal rainfall variation in Malaya, *Annals of the Association of American Geographers*, **58**, 320–326.

Nieuwolt, S., 1971, The 1969/70 rainfall season in the Lusaka region, Zambia, *East African Geogr. Review*, **9**, 11–24.

Nieuwolt, S., 1973a, *Rainfall and Evaporation in Tanzania*, BRALUP Research Paper No. 24, Dar es Salaam, pp. 8, 35–46.

Nieuwolt, S., 1973b, *The influence of aspect and elevation on daily rainfall: some examples from Tanzania*, paper presented at Technical Conference on the agroclimatology of the highlands of Eastern Africa, Nairobi.

Nieuwolt, S., 1974a, Seasonal rainfall distribution in Tanzania and its cartographic representation, *Erdkunde*, **28**, 186–194.

Nieuwolt, S., 1974b, Rainstorm distributions in Tanzania, *Geografiska Annaler*, Series A, **56**, 241–250.

Nieuwolt, S., 1974c, *Interdiurnal Rainfall Variability in Tanzania*, paper presented at International Tropical Meteorology Meeting, Nairobi, preprints pp. 252–257.

Orchard, A. Q., and Sumner, G. N., 1970, *Network Report No. 4, East African Rainfall Project*, London, 58 pp.

Orchard, A. Q., and Sumner, G. N., 1972, *Interim Report, East African Rainfall Project*, London, 50 pp.

Petterssen, S., 1969, *Introduction to Meteorology*, 3rd ed., Tokyo, McGraw-Hill, p. 262.

Ramage, C. S., 1952, Diurnal variation of summer rainfall over East China, Korea and Japan, *Journal of Meteorology*, **9**, 83.

Ramage, C. S., 1964, Diurnal variation of summer rainfall of Malaya, *Journal of Tropical Geography*, **19**, 66–67.

Rapp, A., Berry, L. and Temple, P. H., 1973, *Studies of soil erosion and sedimentation in Tanzania*, (Eds.), Research Monograph No. 1, BRALUP, Dar es Salaam, 274 pp.

Riehl, H., 1954, *Tropical Meteorology*, New York, McGraw-Hill, 392 pp.

Sansom, H. W., 1965, *The maximum possible rainfall in East Africa*, E.A.M.D., Nairobi, Techn. Memorandum No. 3, 17 pp.

Sellers, W. D., 1965, *Physical Climatology*, Chicago, University of Chicago Press, p. 5.

Sellick, N. P., 1953, *Intensity–duration curves for rainfall in Rhodesia*, Meteorological Notes, Series A, No. 2, Salisbury, Rhodesia, 9 pp.

Steinhauser, F., 1966, Über den Tagesgang des Niederschlags, *Archiv f. Met., Geophysik und Bioklimatologie*, Vienna, Series B, **145**, 1–35.

Taylor, C. M., 1968, *Rainfall frequency/intensity data for Kenya, Tanzania and Uganda*, Nairobi, E.A.M.D., 15 pp.

Taylor, C. M. and Lawes, E. F., 1971, *Rainfall intensity–duration–frequency data for stations in East Africa*, E.A.M.D., Techn. Memo. No. 17, Nairobi, 30 pp.

Thompson, B. W., 1965, *The diurnal variation of precipitation in British East Africa*, E.A.M.D., Techn. Memorandum No. 8, 70 pp.

Thornthwaite, C. W., 1948, An approach toward a rational classification of climate, *Geogr. Review*, **38**, 55–94.

Torrance, J. D., 1959, *The incidence of dry spells during the rainy season in Rhodesia and Nyasaland*, Salisbury, 11 pp.

Troll, C., 1964, Karte der Jahreszeitenklimate der Erde, *Erdkunde*, **18**, 5–28.

Watts, I. E. M., 1954, Line squalls of Malaya, *Malayan Journal of Tropical Geography*, **3**, 1–14.

Watts, I. E. M., 1955a, *Equatorial weather*, London, University of London Press, pp. 67–71.

Watts, I. E. M., 1955b, *Rainfall on Singapore Island*, Journal of Tropical Geography, **7**, 43–53.

Weischet, W., 1965, Der tropisch-konvektive und der aussertropischadvektive Typ der vertikalen Niederschlagsverteilung, *Erdkunde*, **19**, 6–14.

Weischet, W., 1969, Klimatologische Regeln zur Vertikalverteilung der Niederschläge in Tropengebirgen, *Die Erde*, **100**, 287–306.

Wexler, R., 1954, *The physics of tropical rain*, Chapter 6 of Riehl, H., *Tropical Meteorology*, McGraw-Hill, pp. 155–176.

Wiesner, C. J., 1970, *Hydrometeorology*, London, Chapman & Hall, p. 6.

WMO, 1971, *CLINO – Climatological normals for climate and climate ship stations the period 1931–1960*, Geneva, No. 117. TP.52, 356 pp.

Woo Kam Seng, S., 1968, *Rainfall in the Highlands of Malaya*, Dissertation, University of Singapore (unpublished), 41 pp.

von Wiszmann, H., 1948, Pflanzenklimatische Grenzen der warmen Tropen, *Erdkunde*, **2**, 81–92.

Wycherley, P. R., 1967, *Rainfall probability tables for Malaysia*, Planting Manual No. 12, Rubber Research Institute of Malaya, Kuala Lumpur, 85. pp.

CHAPTER 9

Tropical Climates

In this chapter a general survey of the various tropical climates of the world will be given in a system of climatic regions with relatively homogeneous conditions. The main features of each climatic type will be explained and the various elements will be described as they synchronize during the course of the year. The treatment and level of generalization are necessarily uneven, because there exist large differences in availability of climatic data and literature between the various parts of the tropics.

Conditions in each climatic region will be illustrated by simple temperature–rainfall diagrams for representative stations, and the scales for these diagrams are, with a few exceptions, uniform, so that a comparison between the various climatic types is easy. Most diagrams are based on data for the period 1931–1960 (Clino, 1971).

As temperature and other climatic elements show a great deal of uniformity in the tropics, subdivisions of the tropical climates have commonly been based on rainfall conditions, in particular the average amount per year and the seasonal distribution (Creutzburg, 1950; Jätzold, 1970; Köppen, 1936; de Martonne, 1941; Thornthwaite, 1948; Troll, 1964; von Wissmann, 1948).

The same basic method will be followed in this chapter. But the various categories will be defined differently in each major part of the tropics, depending on the main factors of origin and the most important characteristics of the seasonal rainfall distribution; and they will be defined on the basis of practicality, that is the creation of a reasonable number of relatively homogeneous climatic regions, each representing a clear type of tropical climate. This pragmatic approach reflects the opinion that a climatic classification which results in a meaningful subdivision in one part of the tropics, is not necessarily the most efficient one in another.

The tropics, as delimited in Chapter 1, occupy four main areas:

(1) Southern and southeastern Asia and northern Australia. This area will be called 'tropical monsoon Asia'.
(2) Large parts of Africa.
(3) Tropical Central and South America, including the Caribbean area.
(4) Tropical areas of the Pacific, Atlantic and Indian Oceans (Figure 1.2).

These four areas form a convenient framework for a survey of the tropical climates.

9.1 Tropical monsoon Asia

This very large area is the home of about 750 million people, or approximately 20 per cent of the total world population. Its main climatic characteristics are almost completely controlled by the monsoons. To be sure, the Asian monsoons also affect large areas outside the tropics, especially in eastern Asia, where their influence reaches as far north as Hokkaido and Sachalin, but these non-tropical areas are not considered here.

Tropical monsoon Asia is best subdivided according to a decisive characteristic of the monsoons, namely whether they bring precipitation or not. This factor controls both the seasonal rainfall distribution and its total amount, the main features that differentiate the climates of this region. But this specific quality of the monsoons is also of extreme practical significance for the inhabitants of the area, dependent as they are on the monsoon rains for their agricultural production. Moreover, most other elements of climate, such as temperature, cloudiness, humidity and local wind systems, are strongly influenced by this same factor. Accordingly, tropical monsoon Asia is subdivided into three climatic types:

(a) *equatorial monsoon climates*, where both monsoons bring rainfall, consequently no regular dry season occurs;
(b) *dry and wet monsoon climates*, where one monsoon brings most of the rainfall, while the other one is relatively dry;
(c) *dry tropics*, where both monsoons bring little or no precipitation.

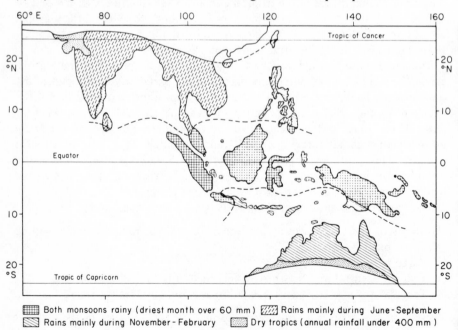

Figure 9.1   The main climatic regions in tropical monsoon Asia

The distribution of these three types shows a clear and relatively simple pattern (Figure 9.1).

9.1.1 Equatorial monsoon climates

In tropical monsoon Asia these climates prevail over a large region between 10° S and 8° N, comprising most of the Indonesian archipelago, Malayasia, New Guinea and some other islands (Figure 9.1). The main characteristic of this region is the mixture of land and sea surfaces, making it truly a 'maritime continent' (Ramage, 1968). Most islands are mountainous and this factor creates a large variety of local climates, depending mainly on exposure and elevation. Nevertheless, a certain homogeneity prevails over the region, because the climates are largely controlled by the monsoons, which here are remarkably similar.

The *north-east* monsoon, which dominates the general circulation from about December to March, gradually changes into a northwesterly wind near the equator (Figure 5.2). South of the equator, and especially in Java, it is often referred to as the 'west' monsoon, or, because it brings a large proportion of the rainfall, as the 'wet' monsoon.

The *south-west* monsoon, which prevails from about June to September, is a continuation of a southeasterly wind in the southern hemisphere (Figure 5.3). In Java it is usually called the 'east' monsoon. It brings rather dry air masses, particularly to eastern Java, and therefore it is locally also known as the 'dry' monsoon.

Generally, however, the two monsoons are similar in that they both bring predominantly warm and humid air masses to the region. These characteristics are related to the high sea surface water temperatures in the region itself and in the oceans around it, which are almost everywhere well above 25 °C. The dense equatorial rainforests on the islands themselves also produce large amounts of water vapour. The air over this region therefore contains more water vapour than over any other equatorial area (Lockwood, 1974).

These very warm and humid air masses need little uplifting to produce large amounts of rainfall; and four main processes cause upward air movements in the region:

(1) *Convergence* may take place either between different branches of the monsoon, or within one monsoon current, when it changes direction or slows down in the vicinity of the equator (Figures 5.2 and 5.3). The zones of convergence are highly variable, both in position and in intensity.

(2) *Convection*, caused by surface heating over land areas during the day, is possibly the most frequent cause of rainfall. Nighttime convection over sea occurs less often but can produce local thunderstorms and disturbances of the 'Sumatra'-type (p. 76).

(3) *Orographic lifting* is another common cause of rainfall in an area with many mountains. Though the monsoon currents in this region are usually not very strong, orographic lifting may locally intensify other rain-producing movements along coastlines and mountain slopes. This factor creates large

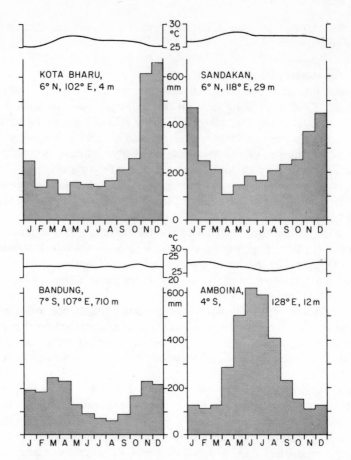

Figure 9.2   Climatic diagrams for four stations in the equatorial monsoon climates of South-East Asia

seasonal variations in rainfall as exposures change with the wind directions of the general circulation.

(4) *Local wind systems* develop very rapidly in this region, where insolation is intensive throughout the year and the general circulation frequently rather weak. By their very nature the local winds influence especially the daily rainfall distribution.

Each of these four processes can produce rainfall on its own; but they usually occur in combination, reinforcing the effect. It is therefore no wonder that this region is one of the rainiest in the world, with annual totals everywhere over 2000 mm (Figure 8.2). However, large differences occur within the region, the extent of which are not well known, as there are few rainfall stations, especially on some of the smaller islands. This is mainly caused by the limited practical importance of these differences in a region where droughts are a rare occurrence.

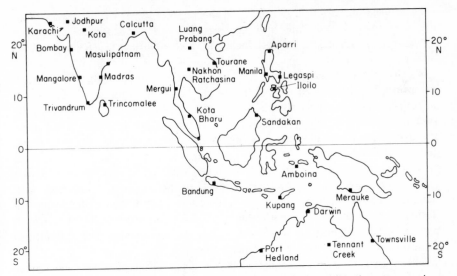

Figure 9.3   Location of stations in tropical monsoon Asia for which diagrams are given

The seasonal rainfall distribution also shows large variations within the region (Figures 9.2 and 9.3). Local factors are the main reason for these differences, principally orographic lifting and local wind systems, which change in importance during the course of the year. Over the region as a whole the intermediate seasons between the monsoons, around March–April and September–October, are probably the wettest. During these periods the doldrum zone, with its widespread convergence, strong convection and active local winds, is over the region and rainfall is very general (Braak, 1950).

Differences in the seasonal rainfall distribution are also of little practical consequence, because dry periods are rarely prolonged, except during particularly dry years (Nieuwolt, 1966).

The predominance of convectional processes is illustrated by the high frequency of thunderstorms in the region (Figure 6.2). Some stations report very high figures, as for instance Bogor with an annual mean of 322 days with thunderstorms per year, which is probably a world record. Over the region as a whole the annual means are much lower, but still impressive. Cumulonimbus clouds are a very common occurrence over the whole region and they reach great heights, up to 15 000 m (Zobel and Cornford, 1966).

Related to the prevalence of convection, rainfall intensity is generally high. The percentage of the total rainfall which is received during cloudbursts (showers of at least 5 mm of rainfall during 5 minutes) varies in Indonesia between 8 per cent and 37 per cent, with an average of 22 per cent. (For comparison: the highest value in Bavaria, Germany, is 3·7 per cent.) At Bogor, over 80 per cent of the annual total rainfall is normally received during showers which bring at least 20 mm (Mohr et al., 1972).

The diurnal distribution of rainfall in this region follows the usual pattern of afternoon maxima over land and night maxima over the sea. Coastal stations show large variations, due to both local and regional influences (Ramage, 1964; Nieuwolt, 1968a).

Because of the similarity of the air masses and the limited seasonal variation in insolation, the other elements of climate all present an impression of *seasonal uniformity*. This is especially true for temperatures: over the whole region the annual range is less than 5 °C (Figure 9.2). Temperature differences with place are also generally small, because compensating winds, such as the sea breeze, develop rapidly. Because of the limited importance of the Coriolis force, pressure gradients remain generally small and wind speeds low (Braak, 1950).

One exception to the uniformity of temperature is its reduction with elevation, a very important feature in a region with many mountains. Many hill stations owe their existence to the relief they can provide from the high physiological temperature conditions in the lowlands and to the agricultural possibilities provided by lower temperatures.

Where seasonal variations are small, diurnal differences assume greater importance. The diurnal range of temperature exceeds the annual range and increases rapidly with distance from the sea, providing a useful indicator of continentality. The diurnal range also increases with elevation, because daytime temperatures decrease much less with height than those during the night.

However, monthly means, indicating seasonal uniformity, and mean daily ranges fail to illustrate the important day to day variations in the weather conditions in this equatorial region. These are frequently related to cloudiness and they affect temperature, humidity and local winds much more than is generally allowed in descriptions of the equatorial monsoon climates (Nieuwolt, 1968b).

Despite many local variations, the region as a whole has numerous common characteristics of climate: plenty of rainfall, high humidity, frequent thunderstorms, many hours of sunshine, dominance of diurnal over seasonal variations and, most important of all, a system of two very similar monsoons (Braak, 1928/29; Nieuwolt, 1969; Sukanto, 1969; Watts, 1955).

9.1.2 Dry and wet monsoon climates

This type of climate prevails in four main regions in tropical monsoon Asia:

(a) the Indian subcontinent, including the northern parts of Sri Lanka and coastal parts of Bangladesh;
(b) Burma, Thailand and Indo-China;
(c) the Philippine archipelago;
(d) northern Australia, southeastern Indonesia and the southern part of New Guinea (Figure 9.1).

The main feature of the climates of these regions is the big difference in character of the monsoon winds. The summer monsoon generally brings warm and humid air masses, which produce large amounts of precipitation, and the winter monsoon is mainly dry. Only a few exposed coastal areas, where the winter monsoon arrives after a long journey over sea, such as southeastern India, northeastern Sri Lanka and the eastern parts of the Philippine islands, experience rainfall during the winter monsoon season (Figure 9.1). In those areas, the summer monsoon is relatively dry and the name of 'wet and dry monsoon climates' therefore remains appropriate.

9.1.2.1 The Indian subcontinent

Conditions in the tropical wet and dry climates of monsoon Asia are best documented in this region, where a uniform system of meteorological observations and publications has existed since the days of British domination. Statistical data for the whole region are therefore comparable and based on these a number of excellent climatological sources are available (Eliot, 1906; Das, 1972; Flohn, 1960; Kendrew, 1937; Lockwood, 1965; Simpson, 1921).

In this region, four seasons can be recognized: the cool season; the hot and dry, or pre-monsoon season; the season of the general rains; and the period of the retreating monsoon.

The *cool season* lasts during January and February in the south, and approximately one month longer in the north of the region. During this period, the winter monsoon prevails over most of tropical India, though in the north it is frequently replaced by winds of a westerly direction.

During this period, temperatures generally decrease with increasing latitude (Kendrew, 1937, p. 118). January means range from about 25 °C in the south to below 20 °C in the northern parts of the region (Figures 9.4 and 9.5). Rainfall is almost everywhere below 25 mm for the two main months of the season, except in the south-east of India and the eastern parts of Sri Lanka (Figure 9.5, Madras, Trincomalee). Here, the north-east monsoon, forced to rise by topography after its journey over the Bay of Bengal, brings rainfall in amounts which increase towards the south. Some rainfall is also received in the northern parts of the region from moving depressions, which are reactivated over the Ganges lowlands (Figure 9.4, Calcutta). However, these disturbances are more efficient as rain-bringers in the areas further to the north, outside the tropical parts of India.

During the *hot, dry season*, which lasts from about March to May, temperatures rapidly increase over the whole region, as the overhead position of the sun moves into the northern hemisphere and days get progressively longer (Figures 9.4 and 9.5). The highest temperatures are reached in the north, where cloudiness is little (Figure 9.4, Kota). In coastal areas, where sea breezes bring cooler air during the hottest hours of the day, and in the south, where cloudiness is higher and some precipitation is received from convectional thunderstorms, the temperatures are somewhat lower (Figure 9.4, Bombay; Figure 9.5, Mangalore, Trivandrum).

Over most of tropical India the total rainfall during this period from March to May remains well below 100 mm. However, the extreme south receives more than double this amount, both from convectional disturbances and from the

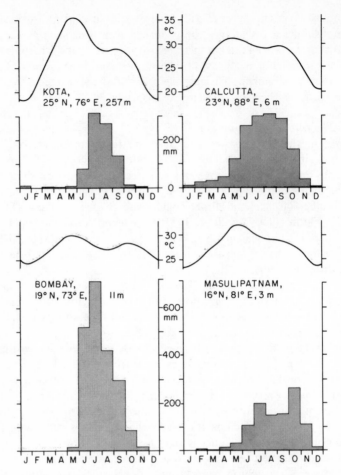

Figure 9.4   Climatic diagrams for four stations in northern India

first beginnings of the monsoon (Figure 9.5, Trivandrum). The north-east of the region also receives more rainfall, caused by local disturbances ('nor'westers') and by tropical cyclones, which originate during the pre-monsoon season over the Bay of Bengal and can bring large amounts of rainfall and heavy damage to the coastal areas of Bangladesh (Chapter 6).

The *season of the general rains* brings over 80 per cent of the total annual rainfall in most of the region. It is entirely due to the summer monsoon. The onset of the monsoon, which is frequently accompanied by violent thunderstorms and therefore often called the 'burst' of the monsoon, progresses from south to north-west (Figure 9.6). With the increased cloudiness temperatures generally are lower than during the pre-monsoon season, creating the 'Ganges-type' of annual curve, shown by most stations in the region (Figures 9.4 and 9.5).

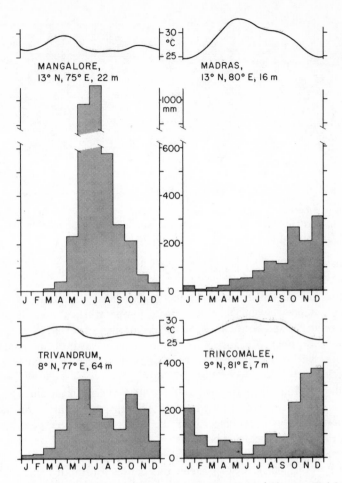

Figure 9.5   Climatic diagrams for four stations in southern India and Sri Lanka

The amount of rainfall during this season varies considerably over the region. Highest totals are reached along the west coast, where orographic lifting takes place (Figure 9.4, Bombay; Figure 9.5, Mangalore). Another area of heavy rainfall is the Ganges delta and the lowlands of Bangladesh, which causes widespread convergence (Figure 9.4, Calcutta). Lowest rainfall totals during this period are found in the central and eastern parts of the Deccan plateau, on the leeward side of the Ghats. After the burst of the monsoon, rainfall decreases here markedly (Figure 9.4, Masulipatnam). The south-east also receives relatively little rainfall during this season (Figure 9.5, Madras, Trincomalee). Another dry area is the north-west where there is a gradual transition to the dry tropics (Figure 9.1).

The *post-monsoon season*, or period of the retreating monsoon, lasts from about October to December, when the south-west monsoon gradually

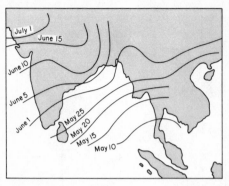 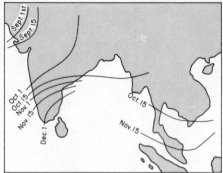

Figure 9.6   Average dates of the onset (left) and withdrawal (right) of the south-west monsoon, and the dates of the start of the north-east monsoon over Indo-China and Malaya (right). (After Das, 1972, pp. 12, 15, and Watts, 1955, p. 9)

withdraws in a southeasterly direction (Figure 9.6). During this period cloudiness decreases and at most stations temperatures show a slight increase, but they remain well below those of the pre-monsoon period (Figure 9.4, Kota, Calcutta, Bombay; Figure 9.5, Mangalore, Trivandrum). Temperature differences within the region are very small, the October mean is everywhere around 26–28 °C (Kendrew, 1937, p. 140).

Rainfall during this period is generally light, except in the extreme south and south-east of the region. Tropical cyclones develop frequently over the southern parts of the Bay of Bengal, where the water surface temperatures reach a maximum in this season. These disturbances generally move westwards and bring variable amounts of rainfall over the coastal areas. In the central and eastern parts of the Deccan plateau this causes a rainfall regime with double maxima at the beginning and end of the monsoon season (Figure 9.4, Masulipatnam).

In this period the winds gradually change to westerly directions in the northern parts of the region, and to northeasterlies in the south, as the winter monsoon gradually establishes itself.

9.1.2.2   Burma, Thailand and Indo-China

In comparison with India, climatic conditions in this region are poorly documented. The number of meteorological stations is very small in some areas, and the accuracy of recordings frequently doubtful; furthermore, no uniformity exists between the different states of this region regarding methods of observation, recording and publication. Records in this region, moreover, have been interrupted at many stations during the Second World War and during internal wars since 1945. Comparison of data for a similar period over the whole region is therefore not possible and publications from many different dates have been used in the compilation of the following description (Air Ministry,

London, 1937; Bruzon, Carton and Romer, 1940; Credner, 1935; Khio-Bonthonn, 1965; Naval Intelligence Division, 1943; Nieuwolt, in press; Staff Members, Academica Sinica, 1957/58; Sternstein, 1962).

As climatic conditions in this region are largely controlled by the same system of monsoons as in the Indian subcontinent, there is a great similarity between the climates of the two regions. However, there are four important differences:

(1) During the *cool season*, the north-east monsoon is much stronger and more continuous over Indochina than over India. It also brings colder and more continental air masses, and consequently the outer limit of the tropics, based on temperatures during the winter, remains well south of the Tropic of Cancer (Figure 9.1). The north-east monsoon is relatively dry, except on the coastal areas of Vietnam, where it has travelled for a considerable distance over the South China Sea with its warm water surface. And a part of the Polar Front

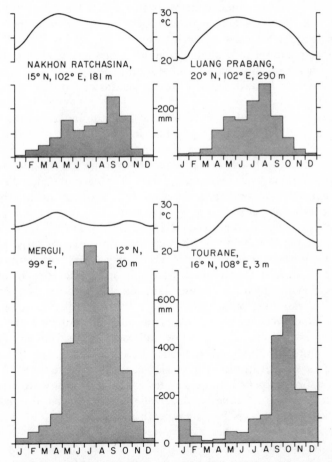

Figure 9.7   Climatic diagrams for four stations in Burma, Thailand and Indo-China.

between the cold continental air masses and warmer oceanic air masses is frequently situated over Indo-China during this period, sometimes as far south as 10 degrees northern latitude (Figure 5.2). Along this front, depressions, which are very similar to the mid-latitude type, frequently bring cool season rainfall to these areas (Figure 9.7). Conditions of heavily overcast skies and drizzle in North Vietnam, known as 'crachin', are related to these disturbances when they have their centre over the Gulf of Tonkin.

(2) During the *pre-monsoon* season, temperatures remain lower than over India. This is mainly due to heavier cloudiness, caused by more frequent convectional thunderstorms probably related to the eastward movement of a trough at the 500 mbar level from India (Ramage, 1955). These bring considerable, though very variable amounts of rainfall over most of the region, the 'mango-rains' of Burma and Thailand (Figure 9.7).

(3) The start of the *general rains* precedes those in India by about 2–3 weeks (Figure 9.6). Also, the onset of the monsoon is more gradual and not so often accompanied by violent thunderstorms. In this region, the summer monsoon is generally weaker than the north-east monsoon, but it is much deeper, reaching up to 9000 m along the coast of Burma.

The proportion of the total annual rainfall which is brought by the south-west monsoon varies from about 80 per cent at stations along the West coast to about 40 per cent on the leeward sides of the mountains and in the interior basins (Figure 9.7).

(4) The *post-monsoon* season is very similar to that in India. But the retreat of the monsoon is earlier over this region than in India, and around November 1st most of it is already under the influence of the north-east monsoon (Figure 9.6). During this season typhoons occur frequently over the South China Sea, where sea surface water temperatures are rather high. The warm and humid air masses, together with the typhoons, produce a rainfall maximum during this period along the coast of Annam (Figure 9.7, Tourane).

Climatic differences within this region are mainly due to relief. The mountain ranges are predominantly in a north-south direction, creating large differences in exposure to both monsoon winds. West-facing slopes receive most rainfall during the south-west monsoon; east-facing slopes, especially in the eastern parts of the region, may receive more rainfall during the cool season (Figure 9.7, Mergui; Tourane). Interior basins are rather dry and frequently show a double rainfall maximum during the beginning and end of the summer monsoon, when wind velocities are still relatively low and local convection therefore more effective (Figure 9.7, Nakhon Ratchasina; Luang Prabang).

Differences in elevations also cause many local variations of climate. Highland stations experience not only lower temperatures, with large diurnal ranges, but also more cloudiness and rainfall than comparable lowland stations. Wind velocities also generally increase with elevation.

In a region which stretches over more than 15 degrees of latitude, some climatic differences are also related to this factor. These are especially evident in cool season temperatures, which decrease with distance from the equator, while the summer temperatures are much more uniform and largely controlled by local factors and cloudiness. Rainfall variability also shows a relation to latitude, it generally increases from south to north, as a larger proportion of total rainfall is caused by travelling disturbances.

Diurnal temperature ranges in this region vary greatly, depending on local relief conditions. They are generally largest during the dry seasons. They also show a clear increase with distance from the sea and give a good indication of the degree of continentality (Nieuwolt, in press, Table 4).

9.1.2.3 The Philippines

This archipelago consists of more than 7000 islands of many different sizes, most of which are mountainous. Local variations of climate are therefore numerous, but certain common features can be recognized, because the general conditions are largely controlled by three major air streams. These prevail during different periods and, accordingly, three seasons can be recognized in this region.

(1) The *north-east monsoon season*, which lasts from about October to March. This wind brings relatively warm air masses to the Philippines and mean sea level temperatures are around 25 °C in most of the region (Figure 9.8). The high temperature of this air, and its high moisture content, is the result of its journey over the warm South China Sea; but this effect is limited to a rather shallow surface layer, reaching usually only up to about 1500 m. Above this layer a weak inversion persists most of the time, and the whole monsoon current is normally not deeper than approximately 2500 m. The north-east monsoon therefore brings much rainfall only when uplifted orographically, as shown by stations with northerly and easterly exposures (Figure 9.8, Aparri; Legaspi). Areas not directly exposed to the north-east monsoon receive only small amounts of rainfall during this period.

The north-east monsoon is overlain by the North Pacific trade winds and the surface position of the convergence during this season is to the east of the Philippines (Figure 5.2).

(2) As the north-east monsoon weakens, this convergence zone moves westwards and during April and May the region is under the influence of the *North Pacific trades*. These bring rather warm, but relatively stable air masses and sea level mean temperatures rise to about 28 °C over the whole region, but rainfall is little except at stations with a direct exposure towards the east (Figure 9.8, Legaspi). The stability of the air masses is related to an inversion layer at approximately 1500 m elevation.

(3) The *south-west monsoon* prevails from about May to September. It is a deep current, up to 10 000 m high, of warm and very humid air masses. These bring large amounts of rainfall, not only to exposed coasts and slopes, but also to most other parts of the region (Figure 9.8). Most of this rain is caused by disturbances which travel with the monsoon. Coastal areas directly exposed to the south-west monsoon have their annual rainfall maximum during this period (Figure 9.8, Manila, Iloilo).

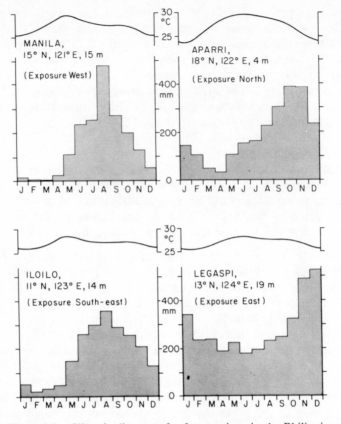

Figure 9.8   Climatic diagrams for four stations in the Philippines

The south-west monsoon reaches its maximum intensity during August. During October, when the south-west monsoon is in retreat and the north-east monsoon not yet well established, southern parts of the Philippines may again temporarily come under the influence of the North Pacific trades.

Another factor causes widespread rainfall, particularly on the northern island of Luzon. This are the typhoons, which occur from June to December, reaching their highest frequency and intensity in September–October, when sea surface temperatures around the Philippines are at their annual maximum. This factor creates a late summer rainfall maximum at many stations (Figure 9.8, Aparri).

As all three air streams will produce rainfall when they are uplifted orographically, exposure is the main factor that controls local climates. Three main types can be recognized:

(a) areas exposed to the south, south-east and west have their rainfall maximum during the south-west monsoon season and are relatively dry during the rest of the year (Figure 9.8, Manila, Iloilo);

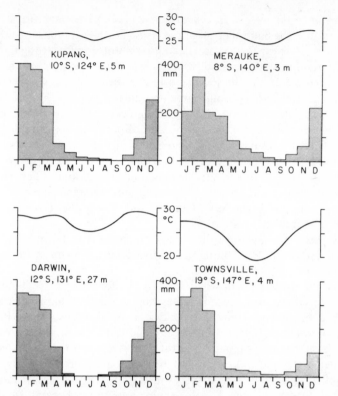

Figure 9.9   Climatic diagrams for two stations in southeastern Indonesia and for two stations in northern Australia

(b) exposure to the north and east produces a maximum of rainfall some time during the period from October to March, but these areas also receive much rainfall in the south-west monsoon season (Figure 9.8, Aparri, Legaspi);
(c) in the southern parts of the Philippines there is a gradual transition to the equatorial monsoon type of climate, which prevails up to a latitude of about 8 degrees North (Flores and Balagot, 1969; Coronas, 1920).

As may be expected, the annual range of temperature shows a clear increase with latitude, and the maxima occur generally during April or May, rather than in mid-summer, when cloudiness is heavier due to the effects of the south-west monsoon.

9.1.2.4   *Northern Australia, southeastern Indonesia and southern New Guinea*
In this region there are two prevailing winds: the west monsoon, usually called 'the monsoon' in northern Australia, and the east monsoon or South Pacific trade wind.

The *season of the west monsoon* lasts from about December to March. The west monsoon is the continuation of the Asian north-east monsoon, after it has changed direction near the equator, and it is reinforced by a strong thermal low over Australia during this period, which is, of course, the southern hemisphere summer (Figure 5.2). As the monsoon has crossed over the warm seas of the Indonesian archipelago, it brings large amounts of rainfall (Figure 9.9). Most rainfall is associated with orographic lifting near the coasts, local convection and two convergence zones, one in the west of northern Australia with south-westerlies from the Indian Ocean, and one in the east with the South Pacific trades. In addition, typhoons in the east and willy-willies in the west of the region bring large, though variable, amounts of rainfall. They reach their maximum frequency in February, which explains the high mean rainfall during that month in southern New Guinea and along the Queensland coast (Figure 9.9, Merauke, Townsville).

During this season, temperature means increase with latitude, from about 26 °C in the north to 28 °C along the northern Australian coast and over 30 °C further inland.

The *east monsoon season*, which lasts from about May to September, when the South Pacific trades prevail, brings hot, dry and stable air masses to the region. These originate from the large dry areas over the southern Pacific just south of the equator (Figure 8.2). The dry and stable characteristics of these air masses are only slightly impaired by their relatively short journey over the sea between Australia and Indonesia, and in the whole region rainfall is very low in this period (Figure 9.9). The dry season lasts between 5 and 6 months and orographic lifting produces only slight amounts of rainfall along the Queensland coast and the southern coast of New Guinea.

During this dry period temperatures show an inverse relationship with latitude, from about 25 °C in the northern parts of the region to below 20 °C near the southern limits.

The homogeneity of this region is therefore one of rainfall, rather than temperatures, which show a strong increase of annual ranges with increasing latitude. The 'Ganges-type' of temperature regime is only represented at a few stations, like Darwin (Figure 9.9).

9.1.3 The dry parts of tropical monsoon Asia
At the outer margins of tropical monsoon Asia, in the extreme north-west and south, are two regions where both monsoons fail to bring rainfall. Total rainfall is normally below 400 mm per year and these regions are therefore classified as parts of the dry tropics (Figure 9.1).

9.1.3.1 Northwestern India
This region, which includes a coastal strip of Pakistan, is dominated from about November to February by westerly winds (Figure 5.2). These bring very little rainfall, not only because they have travelled for a long distance over continental areas, but also because they are warmed and stabilized by their descent from the highlands of Iran and western Pakistan (Figure 9.10; note that the vertical

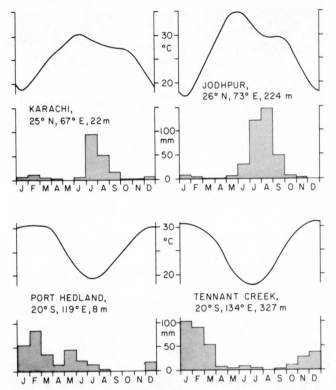

Figure 9.10 Climatic diagrams for two stations in northwestern India and for two stations in tropical Australia. N.B. The rainfall scales on these diagrams are double the scale used in the diagrams for other stations

rainfall scale is double the one used in the other, similar diagrams).

The pre-monsoon season, from about March to June, brings a rapid rise in surface temperatures, which reach impressive heights (Figure 9.10). They cause the build-up of a thermal low, which causes strong sea breezes in the coastal areas. These bring, however, little rainfall as the air masses involved are mostly continental in origin, coming from the north-west and from the Arabian peninsula, reaching this region after a relatively short journey over the sea.

The summer monsoon season starts in this region around July 15th. The main monsoon current comes from the south-east, and though it has lost some of its moisture during the long journey over the Ganges plains, it can nevertheless bring some rainfall to the region; but the actual amounts decrease rapidly towards the north-west (Figure 9.10). However, for most of this season continental air masses from the north-west prevail, and even their convergence with the monsoon air masses fails to produce much rain because the continental air masses are hotter and they are uplifted while the monsoon air remains largely at the same level (Figure 9.11). However, cloudiness may result and surface temperatures are considerably lower compared to the pre-monsoon season (Figure 9.10).

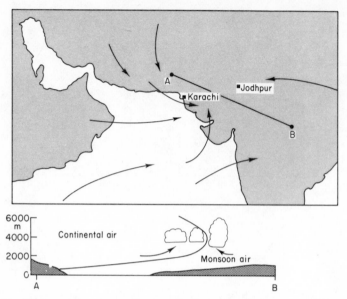

Figure 9.11   Major air streams affecting northwestern India during summer (top). Cross-section along the line A–B (bottom)

In northwestern India the monsoon season ends around September 15th and the post-monsoon period brings a rapid return to winter conditions, with a decrease in temperatures and little or no rainfall.

In this region the annual temperature curves show a modified type of Ganges-regime, combined with large annual ranges (Figure 9.10). The annual ranges, over 17 °C at Jodhpur, are caused by the high latitudinal position, continentality (see the difference between Karachi, on the coast, and Jodhpur, about 600 km inland), and the predominance of clear weather during the pre-monsoon season.

9.1.3.2   Australia

The dry tropics in Australia form a rather narrow transition zone between humid northern Australia and the desert of the interior continent (Figure 9.1).

There are three main reasons why rainfall decreases towards the south. The first one is continentality, causing loss of moisture from the west monsoon as it moves over land. The second reason is the frequent replacement of the west monsoon by southwesterlies from the Indian Ocean. These bring predominantly stable air masses to the region, because they have been cooled by the relatively cold surface temperatures of the West Australian current. This is the reason why the dry tropics extend to the west coast, while the east coast is wetter, as it receives some rainfall from the South Pacific trades.

The third reason for the dryness of this region is that typhoons and willy-willies usually weaken very rapidly when they move inland. They may bring some rainfall in the western coastal areas (Figure 9.10; compare Port Hedland and Tennant Creek).

As in all dry climates, annual temperature ranges are large, considering that the latitude is only 20 degrees South. They increase from the coast, where they are around 10–12 °C, to the interior where they reach values up to about 15 °C (Figure 9.10).

9.2   Tropical Africa

Africa is the largest continent of the tropics and the equator runs through its centre. Tropical Africa, with a population around 275 million, is second in importance amongst the main tropical regions, exceeded only by tropical monsoon Asia.

However, climatic conditions differ from those in the Asian region in a number of important ways. Firstly, there is no general system of monsoons which controls climates over the whole region. Instead, in Africa there are two monsoonal systems, one in the west and one in the eastern parts and they differ in a number of important characteristics (p. 57). And large areas in tropical Africa, like the Congo Basin, the southern parts and Madagascar, have no monsoons but the more continuous winds of the general circulation.

A second set of climatic differences between tropical monsoon Asia and tropical Africa is related to surface features. Tropical Africa is more continental and its coastline is smooth and unindented. Mountain ranges are few, and a large part of the interior constists of extensive highland plateaux, generally at elevations over 1000 m (Figure 9.12). Therefore most of tropical Africa has climates of the continental highland variety, a type that is absent in most of tropical monsoon Asia.

With the equator in the centre of the continent, and its relatively uniform surface, Africa conforms best of all continents to the pattern of climatic types as imagined on a theoretical continent in the tropics (Flohn, 1950; Thornthwaite, 1943).

A third source of differences is the ocean surface temperature. While around tropical monsoon Asia all oceans are relatively warm, tropical Africa is bordered by some rather cool seas. In the west, these are caused by the cold Canaries and Benguela Currents; the latter, in particular, penetrates almost to the equator. In the east, upwelling of cold water from great depths creates low surface water temperatures along the Somalian coast during the period from March to September.

Because of these three types of differences, the subdivision of tropical climates used for monsoon Asia, based on monsoon characteristics, is quite unsuitable for tropical Africa.

Tropical Africa poses another problem: it consists of territories of no less than 37 different states. Collecting comparable data for the whole region is therefore difficult, and this has been the reason why national boundaries are often used in detailed descriptions of the climates of Africa (Griffiths, 1972, p. 15). Fortunately, a number of recent studies and atlases cover the region as a whole and, despite many insufficiencies in the network of meteorological stations, enough data are now available to enable a climatic subdivision based on climatic boundaries to be attempted (Griffiths, 1972; Jackson, 1961; Lebedev, 1970; Thompson, 1965). Griffiths (1972), in particular, contains numerous references, which will therefore not be repeated here.

In tropical Africa, there is a clear tendency for a zone of maximum rainfall to follow the overhead position of the sun with a delay of about one month. This tendency is particularly well developed over East Africa, where the seasonal movements of the I.T.C.Z. are large (Figure 8.5). It is less clear in central and western Africa, where it is limited to a widening and contracting of a zone of heavy rainfall centred on the equator. Nevertheless, the seasonal rainfall distribution over tropical Africa shows a definite zonation, with a double maximum near the equator, and a single maximum at latitudes over about 10 degrees (Griffiths, 1972, p. 27). The single maximum occurs, as a rule, during the period when the sun occupies a relatively high position in the sky.

However, from a practical viewpoint, the total length of the rainy season is more important than whether it occurs in one or two periods of maximum rainfall. The duration of rainy periods has a strong influence on the natural vegetation in Africa (Miller, 1971). It also controls the agricultural growing season and thereby the opportunities facing the farmer in his selection of crops. In large parts of tropical Africa the rainfall in concentrated into a relatively short part of the year, and the length of the rainy periods therefore gives a good indication of the average annual rainfall (Griffiths, 1972, p. 25).

For these reasons the length of the rainy period is used here as a climatic subdivision of tropical Africa. It is indicated by the number of months with a mean rainfall of over 50 mm. This limit constitutes the approximate minimum of rainfall necessary for the cultivation of most tropical crops (Figure 9.13).

It should be mentioned here, that the climatic regions in tropical Africa, as indicated on the map, must be considered as generalizations. They are separated by large transition zones, which belong to one region or the other, according to rainfall differences from year to year. These cause large displacements of the various zones, and the illustrated distribution is rarely realized exactly during any one year.

The distribution of the climatic regions, based on the length of the rainy season, shows a certain amount of symmetry about the equator. The following five main regions can be distinguished:

(1) The *equatorial zone* which has at least 10 humid months. In this region the dry season is absent, or so short that it is of little consequence.

From this central zone, three regions which display a gradual decrease in the duration of the rainy season extend over the rest of tropical Africa.

(2) In *West Africa and the southern Sudan* this decrease progresses from south to north.

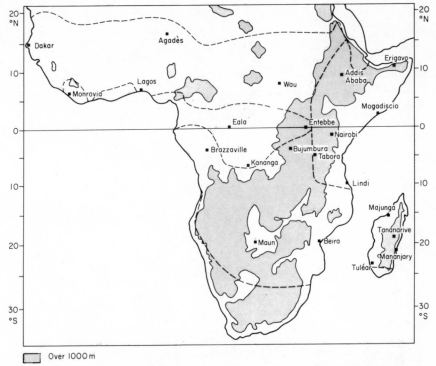

Figure 9.12 Location of stations in tropical Africa for which diagrams are given. Broken lines indicate the approximate boundaries of the main climatic regions. Shaded areas are over 1000 m above sea level

(3) In *Southern tropical Africa*, to the south of the equatorial zone, the duration of the rainy season diminishes from north to south and also from east to west.
(4) In *East Africa* the general decrease is from west to east, but the highlands produce a number of regional climatic variations.
(5) The island of *Madagascar* has relief features which create a quite different distribution pattern; it will therefore be treated as a separate unit (Figure 9.13).

Within each region, many climatic variations are caused by continentality and elevation. Because large parts of tropical Africa consist of highland plateaux, these two factors often combine to influence climatic conditions and their effects cannot be separated easily. Generally, however, all areas at elevations over about 1000 m above sea level show a strong effect of height. The main differences with the adjacent lowland climates are: a reduction of temperature, combined with an increase in the diurnal range, as temperatures decrease much more with elevation during the day than at night; an increase in precipitation and cloudiness, and a higher relative humidity. Because the evaporation and transpiration needs of the vegetation rapidly diminish with height, the natural vegetation is usually more luxurious than in the lowlands. This type of climatic variation occurs in all major regions of tropical Africa (Figure 9.12).

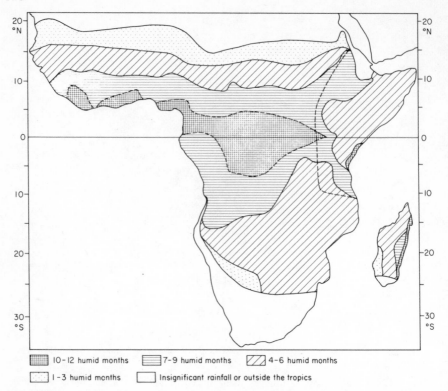

Figure 9.13 The duration of the rainy season in tropical Africa (humid months are those with a mean rainfall over 50 mm). (After Thompson, 1965, p. 54) Thick broken lines indicate boundaries of the main climatic regions

Continentality also effects temperatures, creating larger diurnal and annual ranges. Other effects of distance from the sea are a reduction in precipitation and humidity.

These variations are much more widespread in tropical Africa than in tropical monsoon Asia. On the other hand, purely local differences in climate are generally of relatively small magnitude in areas of uniform relief, such as prevail over large parts of tropical Africa.

9.2.1 Equatorial Africa

This region consists of two main areas: the Congo Basin, between about 5 degrees northern and 5 degrees southern latitude, extending as far east as Lake Victoria; and the southern coast of West Africa, at latitudes between approximately 5 degrees and 9 degrees North (Figure 9.13). In this region, the climate is characterized by heavy rainfall throughout the year, and the annual mean everywhere exceeds 1500 mm, reaching much higher values in places where orographic lifting takes place.

The natural vegetation consists mainly of tropical rainforests of very luxuriant types, which show little or no seasonal variation and are evergreen. The dry season, by definition limited to a maximum of about two months, creates no moisture stress on the vegetation, as it is amply supplied at all times by moisture from the soil. The dry period, where it occurs, is therefore of little importance.

The large amounts of water vapour necessary to maintain this continuous rainfall originate mainly in the southern Atlantic Ocean. However, the area adjacent to the Congo Basin is frequently cooled by the Benguela Current from the south, and the supply of water vapour is therefore limited. Consequently, total amounts of rainfall in the Congo Basin are relatively modest compared with other equatorial regions (Figure 8.2). There is also a large source of water vapour in the region itself, provided by the strong transpiration from the

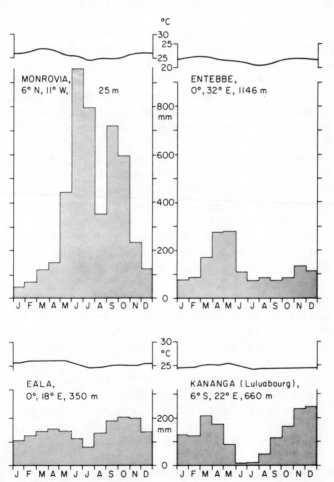

Figure 9.14   Climatic diagrams for four stations in equatorial Africa

luxuriant natural vegetation, supplied by numerous rivers and swamps as well as by rainfall. But this rapid turnover of moisture between the earth's surface and the lower troposphere is not sufficient to maintain the high rainfall totals, as is demonstrated by the decrease of amounts with distance from the west coast (Figure 9.14, Eala, Entebbe).

Along the southern coast of West Africa the water vapour is carried from the ocean by southerly winds, which prevail here during a large part of the year (Figures 5.5 and 5.6). The two interruptions of the continuously wet zone, in western Nigeria and Ivory Coast, are caused by these winds blowing parallel to the coastline, resulting in a divergence caused by differences in friction. Other factors that may be responsible for the dryness in these two areas are the concave form of the coast, causing a divergence of sea breezes, and the presence of very cool water in the Atlantic just south of the coast during July and August, when the Benguela Current may penetrate to areas near the equator (Ilesanmi, 1972; Trewartha, 1962).

Though the main process causing precipitation in this region is convection, seasonal variations are mainly the result of large-scale air-mass movements. Over the Congo Basin convergences between air masses from west and east cause maxima during April/May and October/November, when the main zone of convergence is over this region (Figure 9.14). Minima of rainfall are during December/January in the north, and during June/July in the south of the region. The borderline between these two regimes is almost exactly at the equator (Griffiths, 1972, p. 288).

Along the West African coastal areas, the seasonal movements of the I.T.C.Z. create two periods of minimum rainfall. The major one is during December/January, when dry northeasterly winds prevail (Figures 5.5 and 9.14, Monrovia). The minor dry period is during July or August when the I.T.C.Z. and its associated zone of high rainfall are far to the north of the coast (Figure 5.6). In most parts of this region this dry period is one of reduced rainfall; but it is particularly pronounced in the two areas mentioned before where local divergences create a stability in the monsoon air masses.

Temperatures in this region show very small annual ranges, which increase with distance from the equator, but remain almost everywhere below 5 °C (Figure 9.14). Daily ranges of temperature are between 5 °C and 10 °C, and they increase with both elevation and distance from the coast. Oppressive conditions of high temperature and humidity during the day are therefore replaced by pleasantly cool nights, except along the coasts. However, in the latter locations a strong sea breeze frequently brings relief during the day.

Mean temperatures decrease with elevation at a rate of about 0·65 °C per 100 m. The highland variation of this climatic type is illustrated by Entebbe, where temperatures are strongly influenced by nearby Lake Victoria (Figure 9.14) (Griffiths, 1972, pp. 259–290).

9.2.2 West Africa and the southern Sudan

This is a very large region, stretching over about 15 degrees of latitude, from around 5 degrees to almost 20 degrees North, and from the Atlantic Ocean in the west to the Ethiopian highlands in the east (Figure 9.13). The common feature of climate is the gradual decrease in the length of the rainy season and in the total amount of rainfall with latitude.

This common characteristic is, of course, caused by the movements of the I.T.C.Z. and its associated zones of rainfall during the course of the year (page 110). During December and January this zone occupies a general position at about 2–5 degrees northern latitude in the west, and farther south over the eastern parts of the continent (Figure 5.5). The region is therefore almost entirely under the influence of dry, continental and relatively stable air masses from the north-east or north. Hardly any rainfall is received during this period, except at a few locations along the south coast of West Africa, where local wind systems bring some precipitation (Figure 9.15, Lagos). During this period the temperatures generally increase from north to south, because of more intensive insolation near the equator. The relative humidity is generally low and dusty conditions are frequent, especially in the northern parts of the region, close to the desert.

From February to June the I.T.C.Z. moves in a northward direction, but this movement is not a continuous one. Rather, it is often interrupted or even temporarily reversed (Griffiths, 1972, pp. 170–171). About 500–1000 km south of the surface position of the convergence zone follows a zone of heavy rainfall,

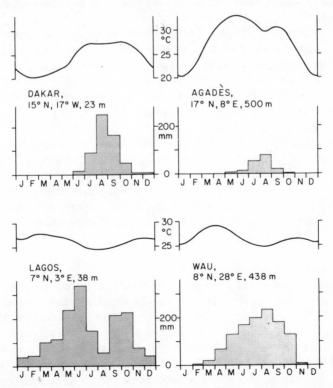

Figure 9.15   Climatic diagrams for four stations in West Africa and the southern Sudan

as the humid air masses of the southern Atlantic Ocean reach a thickness of about 4000 m (Figure 5.7). Rainfall is concentrated along disturbance lines, but convectional thunderstorms also frequently occur in a scattered distribution pattern. In the southern parts of the region rainfall starts in March or April, but in the north not until June (Figure 9.15). During this season, temperatures are generally higher in the northern parts of the region than in the south, a condition that prevails until the end of the northern hemisphere summer, around September or October.

In July and August the I.T.C.Z. is at its most northerly position, at about 15–20 degrees North (Figure 5.6). Rainfall is received over most of the region, except in the coastal areas south of about 8 degrees North where, because of the long distance from the zone of maximum rainfall, a short dry season prevails (Figure 9.15, Lagos). This interruption of the rainy season is, at many stations, not very well shown by mean rainfall figures, because it does not occur during every year in the same month. Over the southern Sudan the convergence of air masses from the south-west and south-east causes much rainfall (Figure 9.15, Wau).

From September to November the I.T.C.Z. moves in a southward direction and its progress is generally a bit faster than during its northward journey. But again the movements are irregular and quite different from year to year. The end of the rainy season comes quite rapidly, during September in the north and during October in the south of the region.

Temperatures in this region show a gradual increase of the annual range with distance from the coast as well as with latitude (Figure 9.15). The seasonal maxima usually come before the start of the rains, and a slight increase after the rains is common, illustrating the effect of cloudiness (Figure 9.15). At Dakar and other stations along the west coast, temperatures are particularly low throughout the year, because the Canaries Current keeps sea surface water temperatures continuously cool.

The northern parts of this region, the Sahel, has suffered a number of exceptionally dry years recently, which caused the decay of vegetation and a progressive southward movement of desert conditions. However, it is too early to decide whether these dry conditions indicate a deterioration of the climate or fall within the normal variation of climate from year to year, in which both wet and dry years tend to show a certain amount of persistence (Griffiths, 1972, pp. 167–186 and 221–231).

9.2.3 Southern tropical Africa

This region covers about 25 degrees of latitude, as it reaches the equator in its northwestern parts and a latitude of 25 degrees South in the south-east (Figure 9.13). Latitudinal differences in climate in this region are therefore of considerable magnitude. As in the analogous region of the northern hemisphere (West Africa and the southern Sudan) these variations are caused by the seasonal movements of the I.T.C.Z.. But in this region these movements differ strongly with longitude: in the western parts the I.T.C.Z. hardly ever moves south of the equator, while in the east it may reach latitudes around 20 degrees South (Figures 5.5 and 5.6). In the southern parts of this region the latitudinal differences of climate are therefore combined with variations between east and west: both the duration of the rainy season and the mean annual rainfall decrease, not only with distance from the equator but also from east to west (Figures 8.2 and 9.13).

The main reason for these differences between the eastern and western parts of southern tropical Africa is the oceanic environment on either side of the continent. In the east, the Mozambique Channel has surface water temperatures which vary between about 23 °C and 27 °C, but in the west the Atlantic Ocean, cooled by the Benguela Current, has surface temperatures around 16–18 °C. The cooling effect of this current is felt almost as far north as the equator and it keeps the I.T.C.Z. from moving southwards during the southern hemisphere summer.

Obviously, the difference in water surface temperatures influences the air masses which move from the oceans to the continent. Generally, these originate from the two subtropical high pressure cells over the Indian and Atlantic Ocean, and they are therefore relatively stable, with an inversion layer at levels between 1000 m and 2500 m. But while the air masses from the Indian Ocean will pick up moisture in their lowest layers from the water surface, those from the Atlantic Ocean will be stabilized by cooling from below. This greatly reduces their ability to produce rainfall when uplifited and the desert areas along the West coast clearly indicate the effects of this factor (Figure 9.13).

Another difference from the northern hemisphere region is that three major air currents meet over this region during the southern hemisphere summer in a triangular zone (Figure 9.16). The southeasterlies bring generally rather dry and stable air masses to this zone, either because they are of continental origin, as in

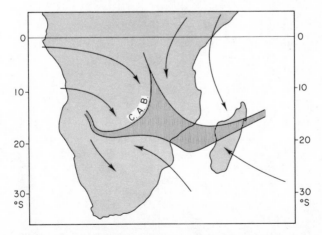

Figure 9.16 Approximate surface position of the I.T.C.Z. (shaded) during December–January. (C.A.B. = Congo Air Boundary)

156

the west, or because they have dropped a large part of the moisture in their lowest layers along the steep eastern slopes of the island of Madagascar. The current from the north-east can be quite humid and unstable, depending on the path it has followed. The air from the north-west is the so-called 'Congo-air', and it is usually quite humid and unstable. This air mass originates over the Atlantic Ocean and has normally picked up large amounts of water vapour over the humid Congo Basin with its luxuriant vegetation, especially along the large rivers. Therefore most rainfall in the southern parts of the region is received along the Congo Air Boundary. In the area of mixed air masses of the I.T.C.Z. itself, widespread rainfall prevails, with local convection or orographic lifting creating strong regional differences and much variation from year to year.

The situation as illustrated lasts during December and January (Figure 9.16). During March and April the whole convergence zone moves northwards and these months bring heavy rainfall in the northern parts of the region (Figure 9.17). The convergence zone leaves the region by the end of May, and the following months, from June to September, are dry. During this period the whole region is under the influence of the southeasterly trade winds. These are frequently of continental origin, coming from the high pressure cell over the southern parts of Africa, and therefore quite dry. But even when they come from the Indian Ocean, they are generally quite stable and have often deposited a large part of the moisture in their lower layers along the steep eastern slopes of

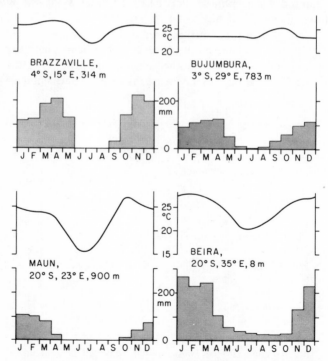

Figure 9.17   Climatic diagrams for four stations in southern tropical Africa

Madagascar, or the coastal mountain ranges. Surface temperatures over the region remain relatively low during this period and therefore convection is not extensive and, as a result of all these factors, this is generally a rather dry season in the region, except along the coast and near mountain ranges where orographic lifting takes place (Figure 9.17). In the western parts of the region the influence of the Benguela Current is particularly strong during this period, as it penetrates almost to the equator. Brazzaville, although more than 350 km from the coast, experiences a drop of temperature of around 3 °C in July (Figure 9.17). Another effect of the cool current is a strong sea breeze, which brings, however, no rainfall.

The southern parts of this region, at latitudes around 20 degrees, experience large seasonal variations of temperature. These generally increase with distance from the coast and with elevation (Figure 9.17, Maun, Beira) (Griffiths, 1972, pp. 232–234 and 409–439).

9.2.4 East Africa

This region occupies an equatorial position, as it reaches from about 11 degrees southern to approximately 14 degrees northern latitude. The common climatic features of this region are two: a relatively low rainfall, compared to other equatorial areas, and a general increase in the duration of the rainy season and the total amount of rainfall from east to west (Figures 8.2 and 9.13). These common characteristics prevail over most of the region, despite numerous local variations, caused by the many mountains.

The relatively low rainfall in a region so close to the equator is related to the monsoonal system, which brings rather dry air masses during a large part of the year (page 59). The main reason for this dryness is the predominantly continental origin of these air masses; and it is intensified by the persistent divergence over most of East Africa during both monsoon seasons, which is caused by a quasi-permanent low pressure centred near Lake Victoria. Other factors which may be of importance are the fact that both monsoon currents run largely parallel to the coastline and the highland edges, rather than across it, the short distance of this region to the high pressure cells in which the monsoons have their origin, the shallowness of the south-west monsoon and the rain shadow effect of the high mountain ranges of the island of Madagascar.

Because of their general dryness, and the frequent presence of an inversion layer in the monsoon air masses, convection is not the major process creating rainfall in this region. These air masses will only produce rainfall after continuous uplifting, and for three reasons this type of movements occurs more often in the western than over the eastern parts of East Africa:

(1) Most mountains and highlands are found in the western areas. The double systems of the Great Rift Valley, with its numerous mountains on both sides, occupies a large part of the western half of East Africa. Orographic lifting will create rainfall here, and many of the mountains are so high that they can be reached by more humid air masses which are occasionally situated over the dry monsoon air.

(2) Air currents from the west create an 'equatorial bridge' type of circulation, which causes widespread convergence (Johnson and Mörth, 1960). These westerly air masses, coming mainly from the Congo Basin, are often rather humid and produce large amounts of rainfall over the western parts of the region.

(3) Lake Victoria, with a surface of over 63 000 square kilometres, and the other large lakes in the western parts of East Africa, produce huge amounts of water vapour and also create local disturbances, as well as the above-mentioned low, all conducive to rainfall.

The general increase of rainfall from east to west is, however, disturbed in a number of places. The first is a rather narrow coastal strip south of the equator, which receives more rainfall than the areas further to the west (Figure 9.13). The main reason is that the south-east monsoon reaches this coast after a relatively long fetch over the Indian Ocean, so that it contains much water vapour in its lowest layers, causing rainfall over a coastal area.

Larger areas where the general rainfall distribution pattern is distorted are caused by relief features. Some mountain ranges in southern and northeastern Tanzania receive more rainfall than the high plateaux further to the west. Another area of high rainfall is around the northern tip of Lake Nyasa, where the lake and the mountain ranges on both sides of it create a strong local convergence during the season of the south-west monsoon. On the other hand, the central parts of the Great Rift Valley, in the rain shadow of the mountain ranges on its eastern edge, are rather dry. Differences in exposure also create large variations in rainfall over relatively short distances. Nevertheless, if the region as a whole is considered, a clear increase of rainfall from east to west prevails (Figure 9.18).

As the monsoons are both relatively dry, the seasonal distribution of rainfall in East Africa is dominated by the movements of the I.T.C.Z. and the distribution closely resembles the theoretical pattern that may be expected over a tropical continent that extends on both sides of the equator (Figure 8.5). The zone of maximum rainfall follows the position of the overhead sun with a time lag of about 4–6 weeks. Accordingly, areas close to the equator have two seasons of maximum rainfall, around April–May and October–November (Figure 9.18, Nairobi; Mogadiscio). This type of seasonal distribution prevails in East Africa between approximately 4 degrees southern and 5 degrees northern latitude (Griffiths, 1972, p. 327). Over most of this area the April–May maximum brings more rainfall, because the I.T.C.Z. is usually very broad in this period and moves only slowly northwards. Locally this season is referred to as the 'long rains'. The October–November maximum, called the 'short rains', generally causes lower amounts of rainfall, as the I.T.C.Z. moves faster on its southward journey, especially over the eastern parts of equatorial East Africa. However, there are many stations where the relative importance of these two periods is reversed, owing to exposure.

Further away from the equator one rainfall season prevails at most stations, around July–September in the northern parts and from about December to

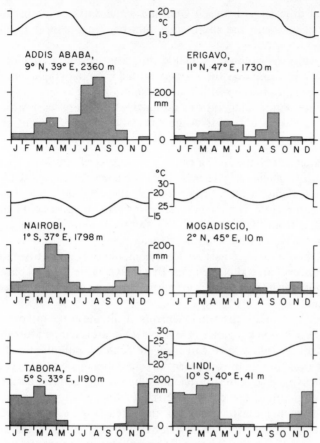

Figure 9.18 Climatic diagrams for six stations in East Africa

February south of the equator. However, the latter season is, at most places, extended to April because the I.T.C.Z. frequently remains present even at 10 degrees southern latitude (Figure 9.18, Lindi; Tabora). There is often an interruption of the rainy season, usually in February or March, but this discontinuity is not very well represented by monthly means because it does not occur every year during the same month (Nieuwolt, 1974).

There are a number of regional departures from this general pattern. The most important is over northern Kenya, Somalia and eastern Ethiopia, where the rainfall maximum of April–July is very small indeed (Figure 9.18; Mogadiscio; Erigavo). The lack of rainfall is caused by a widespread divergence in the south-west monsoon when it is north of the equator. This divergence is the result of four factors:

(a) increase in the speed of the south-west monsoon, which comes more and more under the influence of the very strong low pressure areas in northern India;

(b) frictional differences between land and water surfaces along the southern Somalian coast, where the south-west monsoon runs almost parallel to the coastline;

(c) the pressure distribution, which during this period frequently shows a secondary low pressure over the strongly heated Ethiopian highlands, creating a pressure divergence in the southwest monsoon;

(d) the surface water temperatures along the Somalian coast which are low during the period from June to August because there is frequent upwelling of cold water from lower levels (Flohn, 1964).

A second feature is the spring rainfall during April and May over Ethiopia and northern Somalia (Figure 9.18, Addis Ababa; Erigavo). It is related to upper air troughs which originate in the mid-latitudes, usually the Mediterranean region. The warm surface air forms an upper cold front with the cold upper air from the higher latitudes and the resultant very steep lapse rate causes numerous and violent thunderstorms.

Rainfall variability over East Africa is high, not only in the total amount, but also in the times of arrival of the rains. In a region where in many parts rainfall is marginal for most forms of agriculture this is a serious drawback for development.

Compared to rainfall, the other elements of climate are of minor importance. As can be expected in a region so close to the equator, temperatures show small annual ranges. Within a few degrees of the equator, minimum and maximum monthly means are mainly related to cloudiness conditions (Figure 9.18, Nairobi). In Mogadiscio, the low temperatures in July and August are caused by the cool water along the coast, referred to above. Further away from the equator, the winter season is clearly indicated (Figure 9.18, Lindi; Tabora; Erigavo). Highest temperatures almost everywhere occur before or after the rains (Figure 9.18, Addis Ababa; Tabora). Diurnal ranges of temperature, which exceed the annual ranges, show a clear increase with distance from the coast and also with elevation (Griffiths, 1972, pp. 313-381; Jätzold, 1970; Johnson and Mörth, 1960).

9.2.5 Madagascar

The climates of this island deserve separate treatment and not just because they are strongly influenced by the oceanic position. They are still affected by some continental controls, most important of all the large seasonal shift of the I.T.C.Z. which reaches a latitude of about 15 degrees south during the period from December to February and is consequently frequently over the northern parts of the island (Figure 5.5). But climatic conditions in Madagascar also possess many special features related to the strong relief differences on the island, which has an almost continuous dorsal mountain range with a height of well over 1200 m, in some places reaching over 2000 m (Figure 9.11). Therefore its climates are influenced by elevation and exposure to an extent which is rare on the African continent. This is clearly demonstrated by the main climatic regions and the strong contrasts between them (Figure 9.13).

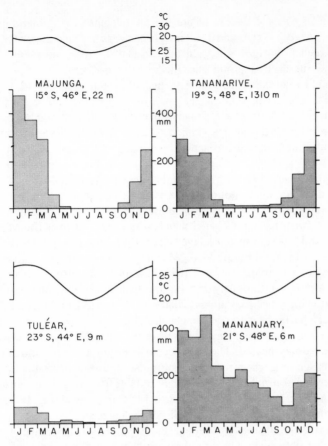

Figure 9.19   Climatic diagrams for four stations in Madagascar

The *east coast*, without a dry season, is almost throughout the year under the influence of the southeasterly trade winds. These are strongly uplifted along the steep eastern slopes of the main mountain range and yield much rainfall, annual means varying between about 2000 mm and 3500 mm (Figure 9.19, Mananjary). The trades are strongest during the winter, when the subtropical high pressure cell at about 30 degrees southern latitude, from which they originate, reach surface pressures up to 1030 mbar. But even more rainfall is received during the summer, especially between January and March, when the trade winds are weaker, but the I.T.C.Z. with its associated depressions and cyclones are over northern Madagascar. These cyclones are formed over the Indian Ocean to the north-east of Madagascar, and they usually recurve to the east of the island, but still bring large amounts of rainfall to the east coast. The relatively low rainfall during September- October is caused by the end of the winter conditions, resulting in a weakening of the trades, coming before the effects of the I.T.C.Z. are felt along the east coast.

The *central parts* of the island are mainly occupied by highland plateaux at elevations between 700 m and 1400 m. These areas are also under the influence of the southeasterly trade winds during most of the year, but these bring much less rainfall here than along the East coast, not only because no further orographic lifting takes place, but also because the lowest, humid, layers of the trade winds are frequently rather shallow. Most rainfall in this region is therefore received during the summer, when the I.T.C.Z. is nearby (Figure 9.19, Tananarive). Large differences related to exposure and elevation occur in this region, and annual means are generally between 1250 and 2000 mm.

The *western areas* of Madagascar show the strongest contrasts between the dry and wet seasons. During the winter these parts of the islands are on the leeward side of the main mountain range and consequently sunny and dry. But from December to March they come under the influence of northwesterlies and the I.T.C.Z. with its depressions, which often develop over the Mozambique Channel, and these areas receive large amounts of rainfall from about November to March (Figure 9.19, Majunga). However, the influence of the I.T.C.Z. weakens rapidly towards the south and the southwestern parts of the island are almost continuously dry (Figure 9.19, Tuléar). Annual rainfall means in this region vary between almost 2000 mm in the north to around 400 mm in the southern parts. Features of the coastal climates in this region are the strong sea and land breezes which develop especially during the summer when the land surface is strongly heated. They also occur during the winter, much more than along the east coast, where the trades suppress almost all diurnal wind systems.

The *extreme south* of Madagascar is outside the tropics, because sea level temperatures during the winter are below 18 °C in some places. It is a semi-arid region with annual rainfall totals well below 400 mm.

Temperature conditions in Madagascar are controlled by three main factors: elevation, latitude and exposure. Elevation causes the largest differences, as means generally decrease by about 0:5 °C per 100 m. Latitude mainly effects the annual ranges, which increase from about 4 °C in the north to approximately 8 °C in the southern parts of the island. Exposure creates a difference between the east and the west, with the west being generally about 1–3 degrees warmer. This difference is especially pronounced during the summer (Figure 9.19, Tuléar; Mananjary) (Griffiths, 1972, pp. 461–484).

9.3 Tropical America

Climatic conditions in tropical America differ from those in monsoon Asia and tropical Africa in two major respects. The first is the very small part of the region which experiences a monsoonal type of circulation. It is limited to a region in South America, east of the Andes mountain range, between latitudes 10 degrees North and 20 degress South. All other parts of tropical America have no monsoonal system, because the I.T.C.Z. makes only rather small seasonal movements. In the western parts these are prevented by the cool

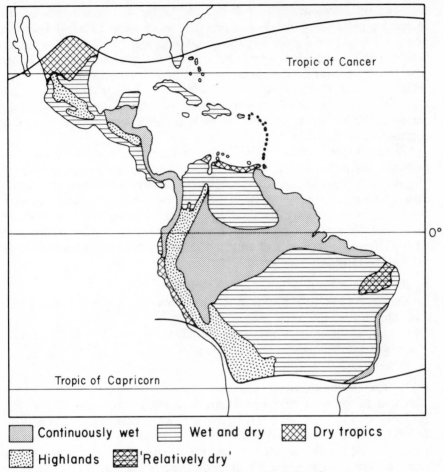

Continuously wet Wet and dry Dry tropics

Highlands 'Relatively dry'

Figure 9.20 Climatic regions in tropical America

Humboldt–Peru Current, which penetrates almost as far as the equator and keeps the I.T.C.Z. from moving to the southern hemisphere. In the eastern parts of tropical America the extent of the South American continent north of the equator is too small to produce a thermal low, and the North American continent is too far away to attract the I.T.C.Z., which therefore remains relatively close to the equator during the summer of the northern hemisphere. Over the Atlantic Ocean east of South America the I.T.C.Z. moves from a few degrees north of the equator to about 3 degrees southern latitude during December to February, but this small movement is not enough to create a monsoonal system. So it is only over the continent itself that the I.T.C.Z. moves far enough south to cause a seasonal reversal of winds (Figure 4.2).

The second characteristic of climates in this region is caused by the long and high Andes mountain range, which reaches to about 50 degrees South and forms a barrier to the strong zonal circulation which prevails over the mid-latitudes of the southern hemisphere. As a result, on the leeward eastern side of the mountains polar air outbreaks can penetrate as far north as about 5 degrees South, where they bring considerable drops in temperature during the winter. Similar polar air invasions also occur in the other tropical regions, but they never reach so close to the equator except in a thoroughly modified form. Another feature related to the presence of the Andes mountain range is that the effects of the cold ocean current and the subtropical high pressure cell over the southern Pacific Ocean, causing dry climates, are limited to a very narrow coastal strip only (Figure 9.20).

The distribution of climatic types, based on the seasonal distribution and total amount of rainfall, show a large number of climatic regions in tropical America (Figure 9.20). Because surface features vary largely in this region, it seems appropriate to divide tropical America into three sub-regions, where climatic conditions can be related to common elements of the general circulation. This is also a suitable arrangement because the relevant literature deals largely with each of these sub-regions, and only in a few cases with tropical America as a whole. The three sub-regions are:

(a) the Caribbean region, including southern Florida;
(b) Central America, the mainland as far south as the border of Panama with Colombia;
(c) tropical South America, to be subdivided into the areas West and East of the Andes and the mountain plateaux themselves.

9.3.1 The Caribbean region

The most important feature of the earth's surface in this region is the large proportion of the area occupied by the sea. The numerous islands and the Florida peninsula are not sufficiently large to create large-scale climatic variations and the climates are therefore predominantly of the marine types. In these latitudes, from about 12 to 27 degrees North, this means that the dominating element of the general circulation are the trade winds, which originate over the Atlantic Ocean. These winds come to the Caribbean area after a journey of about 4000 miles over the tropical Atlantic Ocean. They make only a small angle with the isobars and are almost exactly easterly in direction. More important, they contain large amounts of water vapour in their lowest layers, that is below the inversion. The Caribbean Sea itself has water temperatures of around 27 °C throughout the year, adding much moisture to the lower trades locally.

As is common with trade winds, the North Atlantic trades are strongest in winter, but climatically the intensity and elevation of the inversion layer is of greater importance than the speed and constancy of the winds. Generally, the inversion weakens during the journey over the tropical Atlantic, and its base level increases; but there are strong seasonal variations. During the winter, the inversion is present during about 80 per cent of all days, and its lower level is about 1000–1500 m at the eastern end of the Caribbean region. But from July to

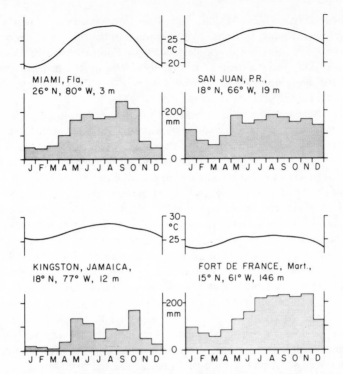

Figure 9.21 Climatic diagrams for four stations in the Caribbean region

October the inversion layer occurs only about 30–40 per cent of all days and its base is about 2000 m or higher, so that rain-bringing cumulus clouds can easily develop in the humid layer below the inversion (Gutnick, 1958). This is the main reason why most of the Caribbean islands show a rainfall maximum during the summer months (Figure 9.21). The main processes causing this rainfall are related to disturbances: easterly waves are most frequent during the months of July to September, and come usually about four to six times per month in that period. From August to October hurricanes also bring much rainfall, though the amounts are highly variable, from place to place and from year to year. The average number of hurricanes in the Caribbean area is about eight per year (Chapter 6).

During the summer there is also some evidence of the I.T.C.Z. being situated over the southern and central parts of the Caribbean region, but its presence seems to have little effect on the rainfall distribution. The zone is frequently rather diffuse and difficult to locate, and there is rarely a strong convergence with southerly winds.

While the seasonal rainfall maximum is generally between June and November, there are many variations within the region. In the southern parts the rainy season may extend up to December, while in the north it ends as early as October (Figure 9.21, Miami; Kingston; Fort de France). In the western parts, there is, at many stations, an interruption of the rainy season during July, similar to conditions along the east coast of Central America (Figure 9.21, Kingston). This interruption is related to a temporary strengthening of the subtropical high pressure cell, caused by northerly winds from the North American continent (De Ward and Brooks, 1934).

As can be expected in a region with many mountainous islands, local variations in rainfall amount and seasonal distribution are quite large. They are related to exposure, with strong differences between west- and east-facing slopes, to local wind systems, especially coastal breezes, and to convection.

During the winter the subtropical high pressure cell over the North Atlantic is rather strong and frequently extends its influence as far south as the Venezuelan coast. The trades are strong and regular, but the persistence of the inversion layer at relatively low levels prevents the formation of rain-bringing clouds, except along strongly exposed slopes. Disturbances are also rare during this season and rainfall is almost at all stations at a minimum (Figure 9.21).

Because of the strong maritime influences, temperatures show small ranges. Annual ranges vary from about 3 °C in the south to about 9 °C in the northern parts of the Caribbean (Figure 9.21). However, these ranges increase with distance from the coast. Some of the low winter temperatures are caused by invasions of cold air masses from the North American continent. These can cause considerable temperature drops in Florida, and smaller ones in Cuba and the Bahamas, occasionally even in Jamaica, but the southern islands are free from this influence (Van Den Berg, 1968; Chang, 1962).

9.3.2 Central America

In its central parts, the American continent consists of a long and rather narrow land mass, which stretches for a distance of over 4000 km in a general north-west–south-east direction. At one place, in Panama, it is only 65 km wide and at two other locations its width is not more than 300 km. This land mass has a high and almost continuous mountain backbone, which broadens in the northern parts (Figure 9.22). This mountain range constitutes a clear climatic divide, so that this region shows strong climatic contrasts between its Atlantic and Pacific sides. The mountains, where they are broad, form a third climatic region (Figure 9.20).

The *Atlantic coastal areas* have climates which are rather similar to those of the Caribbean islands, because they are also under the influence of the North Atlantic trade winds during most of the year. However, the trades are changed somewhat during their trip over the Caribbean Sea and its islands. Most important, the base level of the inversion layer has frequently been lifted and the inversion itself weakened by turbulence and mixing with the humid lower layers of the trade winds. In the lowest layers themselves, more water vapour has been picked up from the warm sea surface. Therefore the trades generally bring more rainfall on the Central American coast than on the islands, and this is particularly the case in the southern parts, where the annual mean rainfall is

Figure 9.22   Location of stations in the Caribbean region and in Central America for which diagrams are given

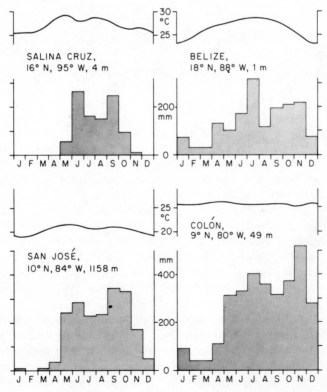

Figure 9.23   Climatic diagrams for four stations in the southern parts of Central America

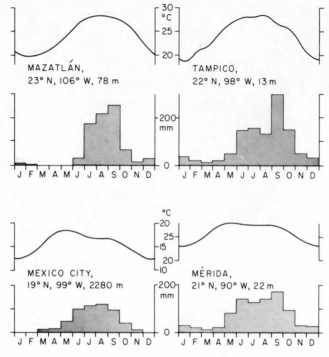

Figure 9.24   Climatic diagrams for four stations in Mexico

almost everywhere over 2000 mm, and even the winter is far from dry (Figure 9.20). As in the Caribbean the season of maximum rainfall is during the summer and autumn (Figure 9.23, Colón; Belize). The reasons are the same as for the Caribbean area: weaker inversion, more disturbances and, in the autumn, hurricanes and the vicinity of the I.T.C.Z.

However, there are three areas of the Atlantic side of Central America where rainfall is lower. One is northeastern Mexico, which lies far to the north of the main tracks of the easterly waves. During the winter this area is also rather dry, because it is frequently invaded by continental air masses from North America. The season of maximum rainfall here is the autumn, when some hurricanes and strong disturbances of various types visit the region (Figure 9.24, Tampico).

A second relatively dry area is the western part of the Yucatan peninsula. The main reasons here are its far northward position and its leeward location in relation to the easterlies (Figure 9.24, Mérida). The third area is the coast of Honduras, which follows an east–west direction and is therefore almost parallel to the prevailing winds, creating a divergence because of differential friction.

In comparison with the Atlantic side, the *Pacific side* of Central America is generally drier. This is particularly the case in the northern parts, where prevailing winds are from the north-west. They bring rather stable air masses from the eastern side of the subtropical high pressure cell over the North Pacific, and these air masses are further stabilized by contact with the rather cool California Current. Only during the late summer and autumn does this region receive more rainfall, as disturbances and even hurricanes occasionally cross over the continent from the eastern side and are reactivated by the high surface temperatures over the coastal lowlands near the Pacific (Figure 9.24, Mazatlán). The number of hurricanes reaching this region varies widely from year to year, but averages about 10 per year.

Summer rainfall along the Pacific side of Central America increases towards the south (Figure 9.23, Salina Cruz; San José). The reasons for this increase are the diminished influence of the California Current as the coast turns more eastward and the higher frequency of winds from the south-west. During the summer and autumn the mountain ranges of the southern parts of Central America often function as the I.T.C.Z., separating easterlies on the Atlantic side from southwesterlies over the Pacific. These southwesterlies bring humid equatorial air masses to these areas, causing large amounts of rainfall when orographic lifting takes place. Summer rainfall at many stations extends into October, because sea surface temperatures over the Pacific adjacent to the Central American coast remain rather high during autumn.

The *mountain highlands* of Central America show a wide variety of climatic conditions, depending largely on elevation and exposure. Most of the more ccentrally located stations are rather dry, not only because of the sheltering effect of mountain ranges on both sides, but also because they frequently reach into the inversion layer of the Atlantic trades (Figure 9.23, Mexico City).

In Central America, annual temperature ranges increase from south to north, with increasing latitude, but also from the coasts to the interior. However, local factors, especially in the mountainous regions, create many departures from this general rule. Diurnal ranges of temperature increase sharply with elevation and distance from the sea (Barrett, 1970; García, 1974; Sapper, 1932; Portig, 1959, 1965).

9.3.3 Tropical South America

The largest part of tropical America is occupied by the South American continent. The most important feature of its surface is the high and continuous mountain backbone of the Andes range. It has already been mentioned that these mountains have important climatological consequences. They divide the tropical parts of the continent into three rather different climatic regions: a narrow coastal strip to the west of the range, the Andes mountain themselves, and the major part of tropical South America, to the east of the mountains (Figure 9.20).

The climates of the *western coastal areas* are controlled by two major elements of the general circulation: the I.T.C.Z. and the subtropical high pressure cell over the adjacent parts of the Pacific Ocean. Because the cold Humboldt–Peru Current keeps the water temperatures of this sea abnormally low for its latitudinal position, almost as far north as the equator, the I.T.C.Z. rarely moves to the southern hemisphere and normally remains throughout the year at around 5–8 degrees northern latitude. The cell of high pressure is located between 10 and 25 degrees South and similarly shows only small seasonal movements.

As a result of these influences, three distinct climatic regions can be recognized along the west coast: a continuously humid region, between approximately 7 and 2 degrees North; a transition zone with rains only during part of the year, between about 2 degrees North and 2 degrees South; and the coastal desert, south of 2 degrees southern latitude (Figure 9.20).

The humid west coast of Colombia is under the influence of the I.T.C.Z. throughout the year. Associated with this zone are westerly winds, mainly from the north-west in the northern parts of the region, and from the south-west in the south. The adjacent part of the Pacific is rather warm, as an equatorial counter-current brings warm water from the west. Therefore the air masses coming to this region are warm and humid, and convergence and orographic lifting result

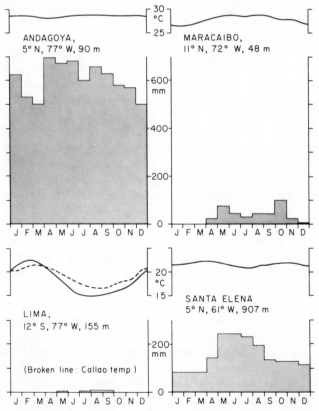

Figure 9.25   Climatic diagrams for four stations in the northern and western parts of tropical South America

in large amounts of rainfall throughout the year, with annual totals everywhere in the region over about 5000 mm (Figure 9.25, Andagoya). Directly along the coast, most of the rain falls during the night, but further inland a clear afternoon maximum prevails.

In the northern parts of the coast the season of lower rainfall is around December–January, when the effects of the subtropical high pressure cell over the northern Pacific are sometimes felt, because it tends to extend its influence towards the equator during this season. In the southern parts of this region the drier season is usually around July–August, when the subtropical high pressure cell over the southern Pacific is strongest.

In the transition zone between 2 degrees North and 2 degrees South the latter influence is of longer duration, and the dry season frequently lasts until December. In this zone large contrasts in rainfall prevail: annual totals to the north of the Gulf of Guayaquil are around 1000 mm, while directly to the south of it they amount to no more than 250 mm. And while the coastal strip has a long dry season, the western mountain slopes of the Andes, 150–200 km inland, receive rainfall throughout the year because of strong orographic lifting (Figure 9.20).

An interesting feature of these regions is that about once in every 11 years the general circulation is completely reversed, as the trade winds from the Caribbean cross over Panama and push the warm ocean currents far to the south. The I.T.C.Z. also moves southwards under these circumstances and may reach a position as far south as 5–7 degrees South. The influence of the subtropical high is weakened and rain may fall over normally very arid parts of the coast. This so-called 'El Niño' effect occurs usually during January or February when the Atlantic trade winds are strongest. It has disastrous effects on the fishing grounds off the west coast as the supply of cold water is temporarily interrupted.

Southwards of about 2 degrees latitude the desert begins. This climate reaches as far as 27 degrees South, well beyond the outer boundary of the tropics, but it is limited to a rather narrow coastal strip (Figures 9.20 and 9.25, Lima. N.B. Lima is actually already outside the tropics, as its winter temperatures are too low, but data for a tropical station in this region were not available).

This region is one of the driest in the world and its extreme aridity is due to the combined effects of two main factors:

(1) The very strong subtropical high pressure cell over the adjacent Pacific Ocean. Air masses from the eastern half of this cell are usually very stable, because they are cooled from below by the cold Humboldt–Peru Current.
(2) The trade winds originating from this cell follow a SSE direction, but are deflected in their lowest layers by the high wall of the Andes mountains. They become almost parallel to the coast, but slightly offshore, and this results in upwelling of cold water directly near the coast, creating very low water temperatures. Frictional differences between land and sea surfaces may also cause some divergence and subsidence, further stabilizing these air masses.

As the relatively warm air masses from the continent get into contact with the very cool water surface along the coast, a low stratus cloud frequently develops. Usually it is a few hundred metres above the surface, as turbulence prevents its formation in the lowest layers; but when conditions are calm the cloud may descend to sea level. This stratus cloud and its associated inversion layer often move inland during the day with the strong sea breezes which occur almost every day in this area. The cloud rarely penetrates far from the coast, as the higher temperatures and numerous thermals over the land rapidly dissolve it. Therefore only a coastal strip, at most locations not broader than a few miles, has a high incidence of low clouds. These clouds occasionally produce a light drizzle, locally called 'garúa'.

Behind the coastal mountain ranges a very dry area follows, where even drizzles are rare. But the western slopes of the main Andes range, which are much higher than the coastal mountains, receive more rainfall, especially during the summer months, when clouds blow over the mountain plateaux from the east.

In this region, temperatures are strongly controlled by distance from the sea. They are continuously low, with small annual and daily ranges, directly along the coast. But even a few miles inland temperature variations are much larger, as is illustrated by the temperature curves for Callao, the harbour of Lima, and Lima Airport, only 5 miles from the coast (Figure 9.25).

The *Andes mountains* have climates which are controlled mainly by elevation and exposure. As in all mountainous regions, local variations are therefore quite large. Elevation has a strong influence on temperatures, but also on rainfall, which generally increases up to a level of about 1000–1500 m, and decreases at greater heights. The interior plateaux usually receive less rainfall than the surrounding mountain ranges. The diurnal distribution of rainfall is also related to elevation: the higher slopes and plateaux, where convection is the main cause of rainfall, have an afternoon maximum, but the lower valleys have most rainfall during the night, when mountain winds converge (Weischet, 1965, 1969).

As easterly winds predominate, east-facing slopes generally receive more rainfall than those facing westerly directions. Similarly the snowline, which near the equator is reached at an elevation of about 4500 m, is lower on the eastern sides of the mountains, because more snow is received there. Exposure is also important in relation to the sun: at higher elevations the temperature differences between sunny and shady areas is large. West-facing slopes, which have more sun during the afternoon, may therefore reach high temperatures than east-facing slopes at the same level.

The seasonal rainfall distribution in this region follows the expected pattern, with the maximum about 1–2 months after the overhead position of the sun. Of the drier periods, the one during June–August is usually the driest, especially south of the equator, because the southeasterly trade winds are rather weak during these months (Figure 9.26, Quito).

Otherwise, the climates of this region have the typical characteristics of tropical mountains: large diurnal temperature ranges, with during the day large differences of temperature between places in the sun and in the shade; generally clear, calm and cold nights, cloudless mornings but afternoons with rapid cloud development over the mountains, frequently strong winds during the day, and a rapid cooling after sunset. The Andes mountains have the highest permanent settlements in the world.

The largest part of tropical South America lies to the *east of the Andes mountains*. It is a predominantly flat area, with a low centre, occupied by the Amazonas Basin, bordered by mountain ranges of moderate height, generally below 1000 m, to the north and south-east (Figure 9.27).

The general circulation over this region is controlled by the position of the I.T.C.Z., of which the seasonal movements are much smaller than over South-East Asia or Africa. This is caused by its relatively fixed position over the adjacent Atlantic Ocean, where it shifts from a few degrees north of the equator in January to about 10 degrees northern latitude in July (Figure 4.2). Over the continent itself the seasonal shift is larger. In July the I.T.C.Z. is a rather diffuse and broad zone situated over the northern parts of the continent at around 7–9 degrees North, but occasionally over the southern parts of the Caribbean area. Most of the South American continent is under the influence of the Atlantic southeasterly trade winds; but in the north these are often replaced by north-easterlies from the Caribbean. In August or September these northeasterlies

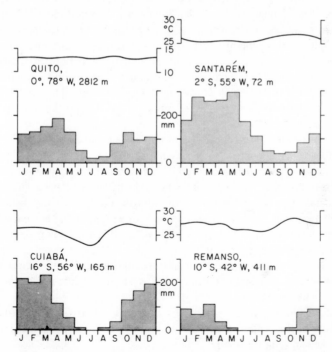

Figure 9.26 Climatic diagrams for four stations in the southern parts of tropical South America

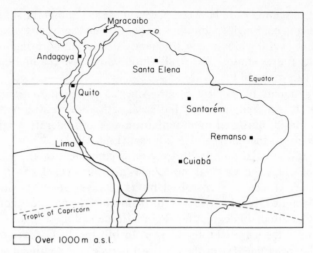

Figure 9.27 Location of stations in tropical South America for which climatic diagrams are given

start to penetrate further southwards. They change into northwesterlies as they cross the equator and normally reach the south-east coast of Brazil, at latitudes between 20 and 25 degrees South, by November. When these winds move over the continent, the air masses pick up large amounts of water vapour and they turn gradually into truly equatorial air masses, rather humid and unstable. This air current and the equatorial air masses prevail over most of Brazil until March. Only the extreme northeastern tip of the country remains under the influence of the southeasterly trade winds from the Atlantic Ocean. The northeasterly air flow retreats around March and is replaced by the southeasterlies in April (Coyle, no date).

The southeasterlies are generally quite dry winds when they reach the South American continent, because they originate from the subtropical high pressure cell over the southern Atlantic, which is quite close. Especially during the period from April to October, the air masses brought by these winds are rather stable. However, as they move towards the equator they gradually become more humid, because they pick up large amounts of water vapour from the luxuriant vegetation cover of the Amazonas Basin, and heating from below rapidly destroys the original inversion layer.

Over the Amazonas Basin and Central Brazil there exists therefore a seasonal reversal of wind directions in a monsoonal system which is not known over the other parts of this region.

The general circulation described above results in two main types of climate in the area east of the Andes: a climate without a dry season and a climate with a clear dry season (Figure 9.20).

The climate without a dry season occupies the equatorial areas of the Amazonas Basin, the north-east coast of Brazil and most of the Guyanas. In this region the total annual rainfall is everywhere over 1500 mm, and the seasonal variation is limited, as there is no real dry season (Figure 9.26, Santarém). The reasons for this continuously high rainfall are related to surface conditions. The many rivers and swamps, with their luxuriant vegetation cover, produce large amounts of water vapour. Local convection rapidly destroys the inversion layers of both the northwesterly and southeasterly air currents, and orographic lifting along the coast and near the Andean foothills produces much rainfall.

The season of maximum precipitation generally corresponds to the overhead position of the sun, with a delay of 1–2 months. However, the period from January to May, when the northwesterly air current prevails, is at most locations wetter than the other period of overhead sun, from August to November, when the I.T.C.Z. moves rather rapidly over this region.

To the north and south of this central zone are located two large areas which have clear wet and dry seasons (Figure 9.20). The wet season is during the period of high position of the sun, therefore during May to August in the northern parts (Figure 9.25, Santa Elena), and from November to March in the areas south of the equator (Figure 9.26, Cuiabá). In areas relatively close to the central zone the dry season is merely a period of reduced rainfall (Santa Elena), but further away from the equator it can be completely dry (Cuiabá). The general pattern therefore indicates a zone of maximum rainfall which moves with the overhead position of the sun. There are, however, three deviations from this scheme in the area to the east of the Andes.

The first is a rather dry zone along the northern coast of Venezuela and including the nearby islands, for instance Curaçao (Figure 9.20). In this area easterly winds from the Atlantic Ocean prevail almost throughout the year. As they blow parallel to the coast, a divergence is caused by differential friction. This divergence is reinforced by acceleration, and the resulting subsidence stabilizes the air masses involved, up to levels of about 2000 m above sea level, high enough to prevent rainfall in most cases. There are only two periods when some rainfall may be expected in this region: during October–December, when the easterlies are relatively weak, and in May–June, when the I.T.C.Z. may occasionally be rather close by (Figure 9.25, Maracaibo).

The second departure from the general pattern is a relatively dry zone in northeastern Brazil (Figure 9.20). This is the area which is not reached by the northwesterly air flow, it is therefore continuously under the influence of the southeasterlies from the Atlantic Ocean. These bring relatively dry and stable air masses from the nearby subtropical high pressure cell over the southern Atlantic. A mountain barrier of moderate elevation prevents the formation of rain during most of the year. The only season of rainfall in this region is from about November to March, when a part of the I.T.C.Z. is situated just south of it (Figure 9.26, Remanso).

Finally, a coastal strip between about 13 and 23 degrees South experiences rainfall throughout the year, caused by orographic lifting of the southeasterly trade winds (Figure 9.20). The season of maximum rainfall in this region is the low-sun period, from about April to September, when the southeasterlies are strongest. During this period mid-latitude cold fronts also cause widespread rainfall, not only along the south-east coast, but also over most of the rest of southern Brazil, up to about 10 degrees southern latitude.

Temperatures in tropical South America to the east of the Andes ranges follow the expected pattern: annual ranges increase with latitude and with distance from the coast, though the latter feature is often obscured by the effects of dry and wet seasons (Figures 9.25 and 9.26). Diurnal ranges show a clear increase with continentality (James, 1939; Serra, 1941; Vulquin, 1971).

9.4   The tropical oceans

The largest part of the tropics is occupied by seas, the tropical areas of the Pacific, Indian and Atlantic Oceans (Figures 1.1 and 1.2). Over these large regions the tropical climates vary much less than over the continents, both from place to place and from time to time. The homogeneous surface conditions reduce climatic differences, even over large distances, and, as water surfaces have a conservative influence, both seasonal and diurnal changes are small compared to those over the continents.

Climatic data obtained over the oceans therefore are representative for quite large areas. Unfortunately such statistics are rare. Data obtained at islands usually reflect local variations rather than the macroclimate, because orographic lifting, coastal breezes and convection affect the climate, even over small islands. Island data are therefore not always reliable indicators of climatic conditions over the oceans. A good example is provided by the data for Honolulu and Hilo in the Hawaiian Islands, which are only about 330 km apart. These two stations present entirely different rainfall figures, and mean temperatures are about 2 °C higher at Honolulu than at Hilo (Figure 9.28). These differences are caused by different exposures: Hilo, on the northeastern coast of Hawaii, gets the full impact of the northeasterlies which prevail in this region throughout the year, while Honolulu is in the south-west of Oahu, on the leeward side of the high mountains of that island.

This is an extreme example, but it shows that climatic data from islands should be treated with caution when used as indicators of conditions over large oceanic areas. On the other hand, they do represent the only areas of human settlements in these regions. Two examples from small islands in the Pacific Ocean illustrate how large differences can be, when distances are measured in thousands of kilometres (Figure 9.28, Truk; Canton Island — distance about 4500 km).

As there are no permanent weather ships stationed in the tropical oceans, the following description uses the generalized world maps of temperature and rainfall given in earlier chapters of this book.

Rainfall differences are of greatest importance, since some parts of the tropical oceans are very dry (Figure 8.2). The main contrast is between the eastern and the western parts of the ocean basins, with the latter always much wetter. The areas with most rainfall are close to the equator, with a slight asymmetry, as they are larger and more intense in the northern hemisphere. This general pattern can, of course, be explained by the main features of the general circulation of the tropical troposphere.

However, there are large regional variations in this pattern. The Indian Ocean shows the effects of the Asian monsoons, and the western parts of the Pacific and Atlantic Oceans differ in the extent of the humid zone, especially in the northern hemisphere. These differences are caused by unequal size and form of the ocean basins, which have a strong effect on ocean currents and related surface water temperatures. The East Asian monsoons also have some influence and all these factors lead to an increased intensity and thickness of the trade winds over the northern Pacific, compared to the northern Atlantic Ocean (Chang, 1962).

The dry areas over the tropical oceans also vary in form and intensity; they are generally larger on the southern hemisphere, where the cold ocean currents are more vigorous and water temperatures lower, than along the corresponding latitudes of the northern hemisphere (Figure 8.2).

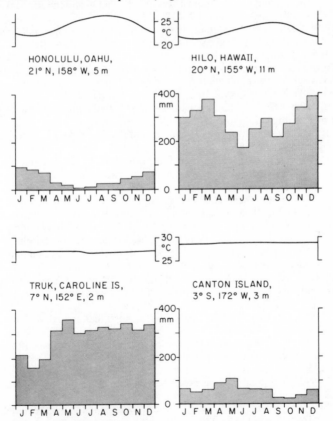

Figure 9.28   Climatic diagrams for four stations in the tropical Pacific Ocean

Temperature conditions over the tropical oceans are characterized by small seasonal variations (Figure 3.2). Surface temperatures are, of course, strongly affected by ocean currents and are lower in the eastern parts of the basins (Figure 3.4). However, these effects are not very clear in the Indian Ocean, where the cool current to the west of Australia is relatively weak (Air Ministry, 1937; Barrett, 1970; Thompson, 1951).

References

Air Ministry, Meteorological Office, 1937, *Weather in the China Seas and in the western Part of the North Pacific Ocean*, Vol. I, 255 pp.; Vol. II, 771 pp., H.M.S.O., London.

Barrett, E. C., 1970, A contribution to the dynamic climatology of the equatorial eastern Pacific and central America, *Transactions, Institute of British Geographers*, No. 50, pp. 25–53.

Braak, C., 1928/1929, *Het klimaat van Nederlandsch Indië*, Vol. I, 528 pp.; Vol. II, 545 pp., Koninklijk Magnetisch en Meteorologisch Observatorium, Batavia.

Braak, C., 1950, Kenmerkende eigenschappen van het tropische klimaat, *Tijdschrift Kon. Ned. Aardr. Genootschap*, **67**, 617–629.

Bruzon, E., P. Carton and A. Romer, 1940, *Le climat de l'Indochine et les typhons de la Mèr de Chine*, Hanoi, Gouvernement générale de l'Indochine, 209 pp.

Chang, Jen-Hu, 1962, Comparative climatology of the tropical western margins of the northern oceans, *Annals, Association of American Geographers*, **52**, 221–227.

Clino, 1971, *Climatological normals for climate and climate ship stations for the period 1931 – 1960*, W.M.O., Geneva, No. 117.TP. 52, 356 pp.

Coronas, J., 1920, *The climate and weather of the Philippines, 1903 & 1918*, Manila, Bureau of Printing, 195 pp.

Coyle, J. R., no date, *Movement of Cold Air Masses over South America*, US Navy reprint, Rio de Janeiro, 8 pp.

Credner, W., 1935, *Siam, das Land der Thai*, Stuttgart, Engelhorn, 422 pp.

Creutzburg, N., 1950, Klima, Klimatypen und Klimakarten, *Petermanns geogr. Mitteilungen*, **94**, 57–69.

Das, P. K., 1972, *The monsoons*, London, Arnold, 162 pp.

De Ward, R. C. and C. F. Brooks, 1934, *Climatology of the West Indies*, in Köppen and Geiger (Eds.), *Handbuch der Klimatologie*, Vol. II,a, Gebr. Bornträger, Berlin.

Eliot, J., 1906, *Climatological Atlas of India*, Bartholomew, Edinburgh.

Flohn, H., 1950, Neue Anschauungen über die allgemeine Zirkulation der Atmosphäre und ihre klimatische Bedeutung, *Erdkunde*, **4**, 161–175.

Flohn, H., 1960, Recent investigations on the mechanism of the summer monsoon of southern and eastern Asia, in S. Basu *et al.* (Ed.), *Symposium on Monsoons of the world*, New Delhi, Hind Union Press, pp. 75–88.

Flohn, H., 1964, On the causes of the aridity of north eastern Africa, translated from *Würzburger Geogr. Arb*, **12**, E.A.M.D., Nairobi, 18 pp.

Flores, J. F. and V. F. Balagot, 1969, *The climate of the Philippines, World Survey of Climatology*, Vol. 8 (Northern and Eastern Asia), Amsterdam — New York, Elsevier, pp. 159–213.

García, E., 1974, Distribución de la precipitación en la república Mexicana, *Boletín del Instituto de geografía, Universidad Nacional Autónoma de México*, **5**, 7–20.

Gutnick, M., 1958, Climatology of the trade-wind inversion in the Caribbean, *Bull. American Met. Soc.*, **39**, 410–420.

Griffiths, J. F., 1972, *Climates of Africa*, Vol. 10 of *World Survey of Climatology* Amsterdam — New York, Elsevier, 604 pp.

Ilesanmi, O. O., 1972, Aspects of the precipitation climatology of the July-August rainfall minimum of southern Nigeria, *Journal of Tropical Geography,* **35**, pp. 55–57.

Jackson, S. P., 1961, *Climatological Atlas of Africa,* Johannesburg, University of the Witwatersrand, 73 pp.

James, P. E., 1939, Air masses and fronts in South America, *Geographical Review,* **29**, 132–134.

Jätzold, R., 1970, Ein Beitrag zur Klassifikation des Agrarklimas der Tropen, *Tübinger Geogr. Studien,* **34**, 57–69.

Johnson, D. H. and H. T. Mörth, 1960, *Forecasting research in East Africa,* in Bargman, D. J. (Ed.), *Tropical Meteorology in Africa,* Nairobi, Munitalp, pp. 56–137.

Kendrew, W. G., 1937, *The Climates of the Continents,* 3rd ed., Oxford, Clarendon Press, pp. 112–151.

Khio Bonthonn, 1965, *Le climat du Cambodge,* Phnom Penh, Service météorologique, 240 pp.

Köppen, W., 1936, *Das geographische System der Klimate,* in *Handbuch der Klimatologie,* Vol. 1, C, Berlin, Gebr. Bornträger, 44 pp.

Lebedev, A. V., 1970, *The Climate of Africa, Vol. I, Air Temperature, Precipitation,* translated from the Russian, Jerusalem, Israel Program for Scientific Translations, 482 pp.

Lockwood, J. G., 1965, The Indian monsoon — a review, *Weather,* **20**, 2–8.

Lockwood, J. G., 1974, *World Climatology, an environmental Approach,* Arnold, London. p. 182.

de Martonne, E., 1941, Nouvelle carte mondiale de l'indice d'aridité, *Météorologie,* **31**, 3–26.

Miller, A. A., 1971, *Climatology,* 9th ed., London, Methuen, p. 127.

Mohr, E. C. J., F. A. Van Baren and J. van Schuylenborgh, 1972, *Tropical Soils,* 3rd ed., The Hague, Mouton, pp. 24–25.

Naval Intelligence Division, 1943, *Indo China,* Geogr. Handbook Series, No. B.R. 510, London, 535 pp.

Nieuwolt, S., 1966, A comparison of rainfall in 1963 and average conditions in Malaya, *Erdkunde,* **20**, 169–181.

Nieuwolt, S., 1968a, Diurnal variation of rainfall in Malaya, *Annals, Ass. of Am. Geogr.,* **58**, 313–326.

Nieuwolt, S., 1968b, Uniformity and variation in an equatorial climate, *Journal of Tropical Geography,* **27**, 25–29.

Nieuwolt, S., 1969, *Klimageographie der malaiischen Halbinsel,* Vol. 2 of *Mainzer Geographische Studien,* Mainz, Geogr. Institut der Johannes Gutenberg — Universität, 152 pp.

Nieuwolt, S., 1974, Seasonal rainfall distribution in Tanzania and its cartographic representation, *Erdkunde,* **28**, 186–194.

Nieuwolt, S., in press, *The climates of continental Southeast Asia,* in *World Survey of Climatology,* Amsterdam — New York, Elsevier, Vol. 9.

Portig, W. H., 1959, Air masses in Central America, *Bull. Am. Meteorol. Soc.,* **49**, 301–304.

Portig, W. H., 1965, Central American rainfall, *Geogr. Review,* **55**, 68–90.

Ramage, C. S., 1955, The cool season tropical disturbances of southeast Asia, *Journal of Meteorology,* **12**, 257.

Ramage, C. S., 1964, Diurnal variation of summer rainfall of Malaya, *Journal of Tropical Geography,* **19**, 62–68.

Ramage, C. S., 1968, Role of a tropical 'maritime continent' in the atmospheric circulation, *Monthly Weather Review,* **96**, 365.

Sapper, K., 1932, *Klimakunde von Mittelamerika,* in Köpper and Geiger (Eds.), *Handbuch der Klimatologie,* Vol. II, h, Berlin, Gebr. Bornträger, 58 pp.

Serra, A. B., 1941, The general circulation over South America, *Bull. Am. Met. Society,* **22**, 173–178.

Simpson, G. C., 1921, The southwest monsoon, *Quarterly Journal of the Royal Met. Society,* **47**, 151–172.

Staff Members, Academica Sinica, 1957/58, On the general circulation over eastern Asian, *Tellus,* **9**, 432–446; **10**, 58–75, 299–312.

Sternstein, L., 1962, *The rainfall of Thailand,* Indiana Un. Research Division, Bloomington, 150 pp.

Sukanto, M., 1969, *Climate of Indonesia,* in *World Survey of Climatology,* Amsterdam—New York, Elsevier, Vol. 8 (Climates of northern and eastern Asia), pp. 215–229.

Thompson, B. W., 1951, An essay on the general circulation of the atmosphere over Southeast Asia and the west Pacific, *Quarterly Journal of the Royal Met. Society,* **77**, 569–597.

Thompson, B. W., 1965, *The Climate of Africa,* Oxford University Press, Nairobi, 132. pp.

Thornthwaite, C. W., 1943, Problems in the classification of climates, *Geogr. Review,* **33**, 233–255.

Thornthwaite, C. W., 1948, An approach toward a rational classification of climate, *Geogr. Review,* **38**, 55–94.

Trewartha, G. T., 1962, *The earth's problem climates,* London, Methuen, pp. 107–110.

Troll, C., 1964, Karte der Jahreszeitenklimate der Erde, *Erdkunde,* **18**, 5–28.

Van Den Berg, C. A., 1968, The Caribbean: Battlefield for weathermen, *Weather,* **23**, 462–468.

Vulquin, A., 1971, Arguments en faveur d'une mousson en Amazonie, *Tellus,* **23**, 1–7.

Watts, I. E. M., 1955, *Equatorial Weather,* London, University of London Press, 224 pp.

Weischet, W., 1965, Der tropisch-konvektive und der ausser-tropisch-advektive Typ der vertikalen Niederschlagsverteilung, *Erdkunde,* **19**, 6–14.

Weischet, W., 1969, Klimatologische Regeln zur Vertikalverteilung der Niederschläge in Tropengebirgen, *Die Erde,* **100**, 287–306.

von Wissmann, H., 1948, Pflanzenklimatische Grenzen der warmen Tropen, *Erdkunde,* **2**, 81–92.

Zobel, R. F. and S. G. Cornford, 1966, Cloud tops over Malaya during the southwest monsoon, *Meteorological Magazine,* **95**, 65–68.

CHAPTER 10

Applied Tropical Climatology

In the field of applied climatology we consider the practical values of climate, its functions in aiding mankind to satisfy his needs, the opportunities it offers and the difficulties it creates. In this respect, tropical climates are important in two ways: as a source of energy and as a major factor in agriculture.

Tropical climates and energy production

One field in which the tropical climates can make significant contributions to the world's economy is in the generation of power. The high amounts of rainfall which are regularly received in many elevated areas of the tropics, constitute a reliable basis for the construction of many hydroelectric power stations. About 55 per cent of the world's total potential of this form of energy is located in the tropics. However, the construction of generators and water reservoirs demands large amounts of capital and many potential sites are situated at long distances from the main areas of power consumption. Therefore only a small proportion of this potential source of power has been developed and of the total hydroelectric power at present generated in the world only 7 per cent has its origin in the tropics.

Solar energy is another source of power in which the tropics are potentially rich, particularly where the seasonal variations in actual hours of sunshine are relatively small. But this natural resource is also undeveloped, owing mainly to technical reasons, and its use is still limited to very small units for domestic purposes. However, once a suitable system of storage has been invented, this form of energy might make a considerable contribution to the economy of the tropics.

The use of wind power, by windmills and sailing vessels, is not basically different in the tropics from other climatic regions. The steadiness of the trade winds, which gave them their name, is an important advantage, and they are still widely used by sailors. On the other hand, many parts of the tropics, particularly near the equator, frequently experience very low wind velocities, so that wind power cannot be used as a constant source. Either an alternative power source or a form of energy storage must be available in these areas, and the lack of either inhibits the use of the wind.

Tropical climates and agriculture

The main function of the tropical climates as a natural resource is in relation to agriculture; and it is a very important one: about 40 per cent of the world's population live in the tropics and all their food and most of their other necessities of life are supplied by agriculture. Because of the low level of industrial development in many tropical countries, the majority of the population is engaged in agricultural production (Table 1). The figures given

Table 1 Percentage of the economically active population engaged in agriculture

Bangladesh — 86	Indonesia — 62	Peru — 62
Bolivia — 72	Laos — 90	Philippines — 66
Honduras — 83	Malawi — 91	Tanzania — 92
India — 72	Pakistan — 75	Thailand — 81

(Source: *UN Statistical Yearbook*, 1972)

here are probably too low, since many people work in their gardens or on small farms in their spare time, and family members frequently help.

In international trade, tropical agriculture is important because it supplies many commodities which cannot be produced in other climates, or which are cheaper to produce in the tropics. Many tropical countries depend on agricultural production for their major export commodities (Table 2).

Table 2 Percentage of total exports contributed by agricultural products

Bolivia — 12	Indonesia — 46	Peru — 28
Colombia — 81	Ivory Coast — 86	Philippines — 73
Honduras — 86	Malawi — 89	Tanzania — 91
India — 31	Pakistan — 90	Thailand — 75

(Source: *UN Yearbook of International Trade Statistics*, 1970)

Frequently, agricultural production in the tropics is a matter of survival for millions of people. Most tropical countries are economically weak and cannot afford food imports on a large scale; they therefore have to rely on food production within the country. Unfortunately, climatic conditions in the tropics can vary widely from year to year, causing large variations in agricultural production. This lack of certainty in a major branch of the economy makes economic planning in tropical countries a difficult task. After a sequence of poor crop years, tropical countries often have to live on the brink of famine in a hand-to-mouth existence until better climatic conditions return.

There is obviously a need to study the relations between climate and agriculture in all climates, but it is particularly important in the tropics. Man cannot change the climatic conditions as they prevail over large areas, but he can adapt his agricultural methods and techniques to these conditions, and he can also change the micro-climate by using irrigation, shelterbelts, shading, mulching or the provision of ventilation or even heating of the lowest layers of the atmosphere, where all plant life takes place. Man can also use the opportunities provided by the climate to a larger extent by the proper selection of crop varieties or the choice of suitable sites, improving drainage conditions, using appropriate techniques of preparing the fields, terracing and transplantation. But maybe the simplest and cheapest way to improve agriculture is to correctly time operations like sowing, weeding, fertilizing and harvesting; for this a good knowledge and understanding of climate in relation to plant life is clearly essential.

As in so many other fields of science, agroclimatology, the study of climate in relation to agriculture, started in the mid-latitudes. However, many methods and conclusions from studies in the mid-latitudes cannot be used in the tropics. Not only are the climatic conditions different, which is obvious, but soil, vegetation, social and economic circumstances tend to pose entirely different questions in the tropics. The largely increased danger and seriousness of soil erosion, the rapid growth of weeds and parasites, the proliferation of diseases, pests and insects are examples of the differences due to climatic conditions. The lack of capital and technical facilities, the poor infrastructure, unreliable statistics and the low educational level of the tropical peasant and his family are all socio-economic conditions that make results of studies in the mid-latitudes irrelevant in most parts of the tropics. The danger of using mid-latitude research and experience without proper adaptation in a tropical environment has been demonstrated by a number of costly mistakes in ambitious schemes, such as the British groundnut development plan in Central Tanzania. Plantations, which are organized and financed by mid-latitude companies, have the means and the attitude to implement research conclusions on a suitable scale, and it is here that tropical agroclimatology has made its major progress. But for the millions of tropical peasants an introduction of new methods, crops or techniques is only possible when all social, economic, cultural and political conditions have been considered.

Unfortunately, climatic conditions in the tropics are generally not as favourable for agriculture as is often assumed. The evidence of a luxuriant vegetation in many parts of the tropics, and the rapid growth of plants, has led to the belief that food and shelter can be produced without great effort. It is a myth, probably kept alive by films and literature depicting idyllic conditions on some tropical islands. In most parts of the tropics, hard and continuous work by the farmer, serious planning and careful timing of agricultural operations are necessary for successful production; and despite these efforts the results are frequently disappointing, mainly because of climatic factors. Four of these are responsible for this state of affairs: they are related to solar radiation, the combination of high temperatures and high humidities, rainfall, and the poor qualities of many tropical soils, caused by climatic conditions.

Solar radiation
The relations between solar radiation and agriculture are primarily in *photosynthesis*. In this process, plants use the visible portion of the solar

radiation to produce carbohydrates (dry matter) in the form of starch, sugar and cellulose. This process is so fundamental, that agriculture has been defined as the exploitation of solar energy. Photosynthesis takes place in the green parts of plants, where chlorophyll pigment is located. Most of the process takes place in the leaves, because these are best exposed to sunlight. The general formula of photosynthesis is

$$CO_2 + H_2O + \text{energy} \longrightarrow (CH_2O) + O_2$$

As sunlight is necessary, photosynthesis is always limited to daylight hours only. The supply both of carbon dioxide and of water can be limiting factors. When these are sufficiently available, the rate of photosynthesis is approximately proportional to the intensity of the insolation, up to a certain maximum, called the saturation light intensity. This maximum varies considerably for different plant species, as some prefer much sunlight while others grow best in shady locations. The maximum rate of photosynthesis also changes with the development of green parts during the growing cycle, especially with the growing of leaves.

The main characteristic of solar radiation in the tropics is the small seasonal variation (Chapter 2). Agriculturally, this is an advantage, as production is possible throughout the year as far as photosynthesis is concerned. In the mid-latitudes, on the other hand, radiation conditions for photosynthesis are generally unfavourable in winter. However, during the summer in these latitudes higher amounts of solar radiation are received than at any time at the equator, because of the combined effect of long days and relatively high position of the sun (Figure 2.4). Therefore, the mid-latitudes are favoured during their agricultural season, which is, of course, all that matters to the farmer. Moreover, the long days of the higher latitudes in summer have the effect of speeding up the growth cycle of many plant and tree species.

Cloudiness reduces the intensity of photosynthesis to a considerable degree, and this effect is rather significant in the humid tropics. It favours the dry climates and the Mediterranean climates during their relatively dry summers. It is also the reason why crops produced during a dry period often have higher yields than those of the rainy season, despite the limitations of water supply.

However, not all the photosynthetic product is retained in the plant. Part of it is used in the process of *respiration*, which follows the general formula

$$CH_2O + O_2 \longrightarrow CO_2 + H_2O$$

Unlike photosynthesis, respiration is a continuous process, which takes place in many parts of the plants. The difference between the gross photosynthesis and the proportion of its products used in respiration, called net photosynthesis, is the rate of accumulation of plant materials in a growing plant.

Respiration increases in intensity with higher temperatures. During the day, when temperature increases are usually the result of more solar radiation, the rates of both respiration and gross photosynthesis go up with temperature and the netphotosynthesis is not greatly affected. But during the night, when

Table 3   Potential net photosynthesis (g/m²/day) in various climates
(Letter symbols refer to climatic types according to the Köppen classification.)

	Humid tropics (Af,Am)	Wet/dry tropics (Aw)	Mediterranean type (Cs)	Mid-lat. humid (Cf,Cw)	High lat. (Df,Dw)
Year	24	25	26	24	18
4 months (summer)	24	27	32	31	33
8 months	24	26	30	29	26

(Source: Chang, 1970, p. 96).

photosynthesis is at a standstill, high temperatures can cause considerable losses of carbohydrates by respiration. This loss is particularly serious in tropical lowlands and coastal areas, because these regions have relatively high nighttime temperatures. The tropical highlands and more continental areas, which generally have larger diurnal ranges of temperature, are much more favoured (Figure 3.5). Outside the tropics, the nighttime temperatures are generally lower; but the main factor which reduces the losses by respiration there are the relatively short nights.

The total effect of these various factors on potential net photosynthesis are summarized in Table 3, which shows the general disadvantage of the tropics in relation to the higher latitudes. The latter are particularly favoured when their crop seasons are compared: for the mid-latitudes this corresponds to about 8 months, while for the higher latitudes, where crops ripen fast during the long summer days, the 4-month period is most relevant. However, one important factor mitigates the disadvantage of the tropics: when there are no other limiting factors, such as rainfall, it is possible to grow two, or even three crops per year in the tropical areas. This possibility exists in many parts of the humid tropics, generally close to the equator. It must also be mentioned that the tropical highlands are insufficiently represented in the above figures, and they are in a much more favourable position than the lowlands as far as net photosynthesis is concerned. This is the result both of more intense solar radiation and relatively low nighttime temperatures (Porter, 1974). Moreover, these highlands usually receive plenty of rainfall or have sources of water for irrigation during dry periods.

The dry tropics are in an excellent position regarding net photosynthesis, and when water is available for irrigation their crop yields are very high (Chang, 1970, pp. 98–100).

It can therefore be concluded that in the tropical lowlands near the equator the best results can be expected from crops with long vegetation periods, such as sugar cane, cocoa, oil palm and rubber. Annual crops will give highest yields where double cropping is possible and in the dry seasons when water is available for irrigation.

Shading is widely practised in the tropics, but it is usually not necessary from the net photosynthesis point of view, except for a few crops which do best at relatively low light intensities, such as coffee and tobacco. The advantages of shading stem from its favourable effects on soil moisture losses, the suppression of weeds and the preservation of soil structure near the surface.

Another effect of solar radiation on plants which can be of importance in tropical agriculture is *photoperiodism*, the response of plant development to the time relations between day and night. Many species will not produce flowers and fruits unless the relative lengths of day and night reach a certain value. In most cases the nights should be longer than the day and plants which pose this condition cannot, therefore, be cultivated near the equator. Most tropical plants are not sensitive to this factor, but some, like rice, sugar cane and coffee, are susceptible to very small variations in the length of the night. These crops can therefore be cultivated successfully close to the equator; their growing cycles are largely governed by photoperiodic factors, even though these vary but little in the course of the year.

Temperature and humidity conditions

Temperatures are, of course, closely correlated with insolation and it is therefore not always possible to separate the effects of these two factors on plant life. However, it is certain that most physical and chemical processes in plants are strongly affected by the temperature conditions. For each plant species there exists an optimum temperature range, at which growth and development proceed with maximum intensity and speed. Every species also has minimum and maximum temperatures, beyond which the plant might be killed or sustain damage. However, these cardinal temperatures differ widely with plant species. In the tropics, where temperatures are uniform and seasonal variations small, the choice of suitable crops in relation to temperature conditions is normally wide. Therefore temperatures are rarely a critical factor in tropical agriculture.

As temperatures in the tropics vary mainly with elevation, it is usually not difficult to determine the correct altitude levels for the cultivation of mid-latitude crops in the tropical highlands. However, the choice of these crops is limited by the condition known as *vernalization*, which means that many species need a cold period before they start flowering and produce fruits. These species cannot therefore, be grown near to the equator. Also, local topography can have strong effects on temperatures, especially the diurnal ranges, so that elevation alone is not always decisive (Kenworthy, 1966).

In the mid-latitudes, low temperatures generally limit the growing season. As temperature conditions are frequently below the optimum for most crops, there exists close correlation between yields and temperature. The system of growing degree-days is based on this correlation. It is also expressed in the van't Hoff law, which states that dry matter production doubles for every rise in temperature of about 10 °C. Such correlations are, however, not valid for many tropical crops. On the contrary, yields of some highland crops, such as pyrethrum, coffee and potatoes, are inversely correlated with temperature (Muturi, 1968). For many other tropical crops, both the quantity and the quality of the products are reduced by high temperatures, mainly because some diseases and pests occur more frequently under warm conditions. The unfavourable effects of high nighttime temperatures on net photosynthesis have already been mentioned.

In comparison with most other climates, relative humidities are generally high in the tropics, and this is certainly the case during the rainy periods (Chapter 7). High humidity of the air has some beneficial effects on plant growth, because many plants can absorb moisture directly from the air and the rate of photosynthesis generally increases with humidity (Baker, 1965). High relative humidities also lower the rate of transpiration, thereby reducing the water requirements of crops, while evaporation losses from the soil are also less than under dry conditions. Generally, crop yields are positively correlated to relative humidity (Arkley, 1963).

The combination of high temperatures and high humidities, that prevails frequently in the tropical lowlands, carries serious drawbacks, however, for tropical agriculture in these areas. It creates highly favourable conditions for the proliferation and growth of numerous micro-organisms and insects, which spread diseases and pests. Weeds and parasites, which can do great damage to crops, also develop more vigorously and more rapidly under these conditions. Losses are not limited to crops on the field; they are equally serious after harvesting and during storage and transporation.

Micro-organisms and pests do, of course, also occur in other climates, but their impact there is much less serious than in the tropics. In the dry climates, the large amounts of sunshine and the relatively cold nights keep their populations well under control. In the mid-latitudes it is mainly the winter, when most of these creatures are killed, that helps to keep their numbers relatively low. During the mid-latitude summers, conditions are rarely as favourable to microbes and insects as in the tropical lowlands, because either humidities or night temperatures are lower. Eradication and control are therefore easy and a matter os seasonal action. But in the tropical lowlands, where this type of climatic restrictions does not occur, the battle against this source of losses is both continuous and costly.

Rainfall

While in the mid-latitudes the growing season and the timing of agricultural activities are determined by the temperature conditions, in the tropics rainfall is the principal controlling element in agriculture. The amount of rainfall that is normally received decides which types of agriculture can be carried out and which crops can be cultivated in a region; the seasonal rainfall distribution regulates the agricultural calendar; and the rainfall variability from year to year is the main factor responsible for fluctuations in yields and total production.

Water is an essential element in plant growth. Its role in photosynthesis has been mentioned and it also acts as the solvent and transporting agent for plant nutrients and provides turgidity in stem and leaves. Water use in plants takes place in the process of *transpiration*, by which water, absorbed by the roots, is transformed into water vapour exhaled by the stomata of the leaves. This process is necessary, not only for the transportation of nutrients and photosynthetic products to all parts of the plant, but also for the cooling of the leaves when these are exposed to the sun for long periods and therefore in danger of being damaged by excessively high temperatures.

Because of these many functions, it is not surprising that lack of water, or moisture stress, reduces the growth and development of plants. Up to a certain maximum, agricultural yields generally show a strong positive correlation with the amount of water available for transpiration. Though this correlation varies quantitatively with the plant species, the stage in the growth cycle and a number of environmental conditions, it remains valid for most crops in the tropics as well as elsewhere. A physical explanation of this relationship is rather complicated, but it has been confirmed by numerous experiments (Chang, 1968, pp. 210–215). This is, of course, why rainfall is the most important factor in agricultural production in many of the drier parts of the tropics.

The relations between rainfall and the water requirements of crops are often illustrated by the *water balance*, which can based on actual data for days, weeks or months, or on monthly means (Thornthwaite, 1948). The general equation of the water balance is:

rainfall = evapotranspiration ± differences in soil moisture + runoff
+ percolation into the subsoil

On the input side, water used for irrigation may be added to the rainfall, and frequently the water balance equation is used to obtain an estimate of the amounts of water needed, and the frequency of irrigation. Evapotranspiration is the combined loss of water by evaporation from the soil surface and by transpiration from plants. Soil moisture, which act as a buffer between rainfall and water use, can be added to the available water for crops, but it can also mean a net loss, when surplus water is stored in the soil for later use. Runoff and percolation to deeper layers of the soil are always water losses.

In water balances, the water requirements of crops are usually indicated by the potential evapotranspiration (E_0), which is the amount that would evaporate from an open water surface, under prevailing conditions. Its value is obtained from evaporation pans, or it is estimated from other meteorological data by various methods (Chapter 7). However, the actual water needs of a crop (Et) may differ from E_0 because the evaporating surface, the total area of the stomata of the leaves, can be different from a water surface. In most cases the total area of the stomata is smaller than that of an open water surface, and the Et/E_0 ratio will be less than $1·0$. But when the vegetation is luxuriant and several layers of leaf canopy are present, as in rain forests, the stomata can transpire more than an open water surface and the ratio will exceed $1·0$.

For crops, the Et/E_0 ratio varies with the type of crop, its stage in the growing cycle and the development of leaves, and the density of the plant cover.

As Et and E_0 are both controlled by the same meteorological factors, their ratio is largely independent of location. Values obtained in one part of the tropics can therefore be used in other tropical areas without serious error. Some

Table 4 Et/E_0 ratios for some tropical crops

Crop	Months after planting						
	1	2	3	4	5	6	7
Groundnuts	0·45	0·80	0·90	0·90	0·90		
Bananas	0·40	0·50	0·60	0·70	0·80	0·90	1·00
Sugar cane	0·30	0:50	0·70	0·90	1·00		
Maize	0·55	1·10	1·20	1·20	1·20		
Sorghum	0·80	1·20	1·20	1·20			
Alfalfa	0·55	1·00	1·00	1·00/cut/			
				0:60	1:00		
Tea (after pruning)	0·00	0·50	0·85				
Eucalyptus	0·40	0·55	0·70	0·80	0·95	1·05	1·30

Seasonal values		Annual values	
Grasses (dry season)	0·81	Evergreen forest	1·00
Grasses (wet season)	0·86	Bamboo forest	0·90
Coffee (dry season)	0·50	Pine forest	0·80–0·86
Coffee (wet season)	0·80	Grasses	0·54 0·85

(Sources: see Nieuwolt, 1973, pp. 7, 14).

values, indicating changes during the growing cycle and estimates for a whole season, are given in Table 4.

Therefore the calculation of crop water requirements for water balance studies creates few practical difficulties. Evapotranspiration values are conservative and vary but little from year to year, so that means over relatively short periods can be used without serious error. And while the Et/E_0 ratio does vary, it remains within narrow boundaries.

The amount of water available to plants during a given month is often different from that supplied by rainfall, because a quantity of water can be stored in the root zone of the soil. A surplus of rainfall over water needs during one month can therefore be used in a following month to make up a deficit. However, the amount of water that can be stored in the soil, the soil moisture holding capacity, varies with the type of soil and the depth of the root zone of the vegetation.

When the water requirements of the vegetation cannot be met by the supply from rainfall and soil moisture, the actual evapotranspiration will be less than the potential one, and the difference between this value and the water need is the water deficit. On the other hand, when rainfall exceeds the water needs of the plant cover, the surplus is first used to recharge the soil moisture to its maximum capacity and then the rest is directly available for runoff, both at the surface and underground.

The water balance method suffers from three main weaknesses. Firstly, the maximum water holding capacity of the soil is rarely well known. There are two main methods of measuring soil moisture, but neither is usually suitable for use by the farmer (Chang, 1968, pp. 194–195). And even in well-conducted experiments, the soil moisture holding capacity is often estimated, because the measuring methods are expensive, time consuming and difficult to interpret. Estimates are based on soil type and depth of the root zone, but local conditions might differ widely from this general value (Thornthwaite and Mather, 1957). The error introduced by this factor is particularly important in climates with pronounced wet and dry seasons, where the vegetation depends on soil moisture for a considerable part of the year. It is therefore a serious disadvantage of the water balance method in many parts of the tropics.

Secondly, the use of mean rainfall figures in water balances is often misleading, because actual amounts of rainfall during an individual year can differ widely from the mean. As an illustration the year 1963 in Malaya may be used, when water deficits of up to 8 months duration and up to 20 per cent of the mean annual rainfall were experienced at stations which were supposed to have only very short and small water deficits (Nieuwolt, 1966). This factor should be considered in many parts of the tropics where rainfall variability is high: water balances based on mean rainfall figures can be unreliable guides in estimating agricultural possibilities. A better way is to use rainfall amounts that can be expected with a certain probability (Nieuwolt, 1973, p. 8).

A third disadvantage of the water balance method is related to the intensity of rainfall. When it exceeds the maximum infiltration speed of the topsoil, a proportion of the rainfall is lost by surface runoff and not available to the vegetation. In the tropics, where rainfall intensities are generally high, this proportion can be considerable, and it is not accounted for in water balance equations.

Despite these three imperfections, the water balance method can provide valuable information to the farmer and the agricultural planner. It indicates the agricultural possibilities of an area, as far as water supply by rainfall is concerned, shows the times and dimensions of water deficits, which allows an estimate of irrigation requirements, and gives an estimate of the right times for the most important agricultural operations.

In the tropics, water balances are characterized by high values of the potential evapotranspiration throughout the year. In the mid-latitudes this factor always shows a clear minimum during the cold season, but in the tropics the seasonal variations of E_0 are generally small. Tropical water balances therefore differ mainly in regard to the rainfall distribution. Three types can be recognized (Figure 10.1). The first is represented by Singapore, where rainfall exceeds E_0 throughout the year. This type occurs mainly in areas close to the equator. There are no regular periods of water deficits and, during occasional dry spells, the vegetation is amply supplied with water from the soil. In these areas, agriculture is rarely limited by rainfall conditions.

The second type of tropical water balance illustrates the other extreme, where

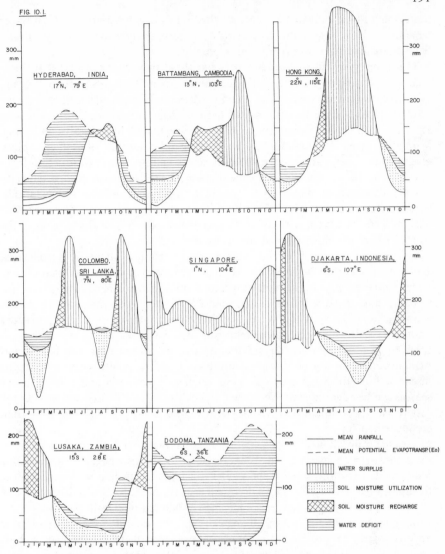

Figure 10.1   Typical water balances at selected stations in the tropics

E_0 exceeds the mean rainfall during all months of the year, as shown by the diagram for Dodoma, Tanzania (Figure 10.1). This type occurs in many parts of the dry tropics. Water surpluses are rare and usually of small size. In these areas crop agriculture is possible, but it is limited to the rainy season and, if no supplementary irrigation is available, to crops which use little water because they can be spaced widely. Maize, groundnuts and many tree crops, such as cashew nuts are grown in these areas, but the main form of agriculture is pastoralism, based on cattle, goats or sheep. In areas where this type of water balance prevails, agricultural activities are completely controlled by rainfall conditions, except where irrigation is possible.

The third type is intermediate between these two extremes and it forms a gradual transition, following the sequence from Colombo, Hong Kong, Djakarta, to Battambang, Lusaka and Hyderabad (Figure 10.1). In that order, water surpluses decrease and water deficits grow in size and duration, and agricultural possibilities are more and more restricted by rainfall conditions. Moreover, as the water balance becomes more unfavourable, rainfall variability generally increases, and with it the differences in agricultural production from year to year. In this third type of water balance the soil moisture storage capacity has the strongest influence on the calculation of water deficits and surpluses, because the vegetation is sustained by soil moisture during at least part of the dry season. The third type, with its many variations, occurs over those parts of the tropics where dry and wet seasons alternate.

Though a dry season usually forces an interruption of many agricultural activities, it can also have beneficial effects. Many tropical crops, such as rice, maize, tobacco, sugar cane, cotton, groundnuts and cocoa need a dry period for ripening and the quality of the products is higher when the last part of the growing cycle falls in a dry season. Yields are often also higher, because photosynthesis is more intensive during dry days, when more sunshine is received than during rainy days. Other crops, like cashew nuts, sunflowers and mangoes, need a dry period for fruiting, as it inhibits the outbreak of fungus diseases. For still other tropical crops, such as pyrethrum and rubber, a dry season can be advantageous because it means a rest period. Production after the rest is usually so much higher that the losses of the dry period are more than compensated for.

Tropical rainfall also carries serious disadvantages for agriculture because of its generally high intensity (p. 122). Not only is the effectiveness of rainfall is reduced, as the infiltration speed of the topsoil is frequently exceeded, but the resulting surface runoff can also do much damage in the form of soil erosion and the washing away of crops, especially newly planted ones with relatively short root systems. Other factors being equal, the extent of soil erosion is approximately proportional to the intensity of rainfall. Therefore, even a short rainstorm can do extensive damage, particularly when it occurs after a dry period, when the vegetation cover is scanty and provides little protection to the soil (Temple and Rapp, 1972; van Rensburg, 1955). Intensive rainstorms, which occur frequently in many parts of the tropics, also have unfavourable effects on soil properties, as will be described in the next section (Nieuwolt, 1974).

Tropical soils
The same climatic factors which affect vegetation and agriculture directly, also influence them indirectly through their role in soil formation. This effect is most important where other soil forming factors, notably parent material and topography, have remained constant over long periods, so that a state of

equilibrium between climate, vegetation and soil has been established. Soils which exhibit such an equilibrium are called zonal soils, and they show the dominating effect of climate, the most important elements being moisture conditions and temperature.

Zonal soils in the tropics are generally rather deep, because the combination of high soil temperatures and high moisture content, which prevails frequently in most parts of the tropics, favours an accelerated chemical disintegration of parent rock material, which proceeds to great depths. The same climatic conditions limit the amout of organic material in tropical soils, because the high intensity of bacterial activity causes a rapid decay of all organic matter and little humus is formed. In this respect a temperature of about 25 °C seems to be a critical value: at lower temperatures there is a general accumulation of humus in the topsoil, but at temperatures above this level, the decay of organic matter proceeds more rapidly than it can be replaced by most types of vegetation (Strahler and Strahler, 1973). Therefore most tropical zonal soils are poor in humus content and their nutrient holding capacity is severely reduced by this condition.

Where rainfall is plentiful, with annual means over about 1500 mm, leaching and eluviation remove most nutrients from the topsoil, and this effect is particularly pronounced where a large proportion of the total rainfall comes in the form of intensive rainstorms, as is the case in most parts of the tropics (Riehl, 1954).

In tropical areas where dry and wet seasons alternate, the evaporation of groundwater near the surface during the dry period may counterbalance the leaching in the rainy season to a certain extent. However, occasionally this process produces a deposit of chemicals near the topsoil in the form of a hard and impenetrable crust, and this effect is most serious near the margins of the dry tropics.

Therefore tropical soils are generally rather unfavourable for agriculture. Fertile soils in the tropics are limited to areas where either the parent material or the topography produces many nutrients. An example of the first case are many volcanic deposits, which weather and disintegrate rapidly under tropical conditions. Favourable topography is found in many river floodplains, where frequent floods leave a deposit of fertile material.

The disadvantages of most tropical soils are aggravated when the natural vegetation cover is destroyed. Under the natural equilibrium of soil, vegetation and climate, a rapid turnover of a relatively small quantity of nutrients takes place in an almost closed cycle between the top layers of the soil and the vegetation. Losses are small, as the soil is well protected against erosion by the vegetation cover, except on rather steep slopes. When the vegetation cover is destroyed, all nutrients in the plants are lost, and the temperature and moisture conditions of the soil are changed. Soil surface temperatures, kept low in the shade of the vegetation, and under the insulating natural mulch cover, can reach very high values in direct sunlight, resulting in a rapid decay of organic matter in the top layers of the soil. At the same time, the direct impact of raindrops can start serious soil erosion and leaching, particularly of nitrogen, calcium, magnesium and potassium, is rapid. A few rainstorms can destroy a soil which took thousands of years to form. To prevent these processes, selective clearing should be practised, making sure that at all times a vegetation cover will be present to protect the soil against direct sunshine and rainfall. The practice of keeping a good undergrowth under many tree crops (rubber, oil palm, cashew nuts) is related to this danger of rapid destruction of the soil when it is bare to the elements of the tropical climate.

The importance of climatic disadvantages

The relative importance of the climatic handicaps to tropical agriculture described in the preceeding section differs widely from area to area. Where rainfall is copious and dry periods usually short, limitations are mainly set by net photosynthesis, pests and diseases and poor soil conditions. Where rainfall is less plentiful, and this is normally combined with higher variability, the importance of rainfall for tropical agriculture increases, while the other disadvantages become less serious. In the drier parts of the tropics, rainfall is the main limiting factor in agriculture, but in the very dry regions it ceases to be of importance, since all agriculture is necessarily based on other sources of water.

In the tropical highlands, all effects of high temperatures are attenuated with increasing elevation. Rainfall, usually rather intensive and variable, is the deciding factor in agriculture. Soil conditions and temperatures also differ strongly with slope and exposure. A rather complicated pattern, related to local topography, of varying conditions exist.

It is one of the main duties of agroclimatologists in the tropics to properly identify the decisive factors of climate regarding agriculture in each area, so that unnecessary efforts of research are prevented.

Man's adaptations to climate

In order to minimize the effects of the climatic disadvantages mentioned above, and to obtain maximum benefits of the opportunities offered by tropical climates, man can adapt his techniques and methods of tropical agriculture.

People living in the tropics have, of course, done this for thousands of years and a large part of most tropical cultures consists of the accumulated experiences of many generations of farmers. Consciously and unconsciously, climatic factors have been taken into account, though usually in a rather unscientific, subjective way. Popular customs and beliefs control tropical peasant agriculture to a large extent and many habits are based on a correct interpretation of climate in relation to plant growth.

In many parts of the tropics farmers try to obtain optimal results by growing a combination of different crops or by practising different types of agriculture at the same time, thereby reducing the effects of failure in any part of their enterprises. Modern science cannot improve very much on this traditional way of spreading the risks in tropical agriculture. A big improvement can be made by the introduction of new crops, or new varieties, often developed in entirely different parts of the tropics, as is demonstrated by the increased yields of new strains of rice and maize.

The tropical farmer can do very little about low net photosynthesis, except allowing his crops more time to grow. In the few cases where sunshine is too abundant, shading is an efficient, cheap and widely used protection.

The effects of high temperatures and humidity, in the form of pests and diseases, can be controlled by modern chemicals. These methods are generally cheap and efficient, but nevertheless many small farmers in the tropics cannot afford them. The selection of disease-resistant crop varieties has helped a great deal to alleviate the damage done by the most serious diseases in cash crops. Tropical diseases in animals are also kept under control by modern chemical and biological methods. Still, many regions have to be avoided, such as the tsetse fly infected areas, where no cattle can be kept. With growing population pressures it will be necessary to combat many diseases of this kind more actively, and methods for this are well known. Their use is still a matter of benefits compared to costs.

The effects of rainfall conditions are widespread, easy to observe and to understand, and therefore many adaptations are available. Where rainfall is scarce, irrigation is the obvious solution. Most damage by lack of rainfall is experienced in the marginal areas, where crop agricultuure is possible without irrigation, but at a high risk. Here, wide spacing of plants or trees will reduce water requirements. Mulching can also have beneficial effects, as it reduces evaporation from the soil.

Where too much rainfall is received, little can be done except to arrange good drainage facilities. The leaching of soils can only be remedied by frequent use of fertilizers. Soil erosion can be prevented or reduced by terracing, contour ploughing, protection of undergrowth and the preservation of a continuous cover of vegetation, so that no bare soil surface is ever exposed to rainstorms.

Because of the high variability of rainfall, there are many parts of the tropics where both droughts and floods can be expected, though they usually do not come during the same year. This combined threat is present in large parts of India and Pakistan, East Africa and Brazil. For economic and technical reasons efficient protection is almost impossible and damage is often rather serious.

The adaptation to climatic conditions is well illustrated by some agricultural systems practised in large parts of the tropics. One of the oldest is 'shifting cultivation', which even today is used, in its various forms, by about 200 million people (Manshard, 1968, pp. 81–100). The common characteristic of all types of shifting cultivation is the frequent movement of the agricultural activities to different plots. An area of land is cleared, cultivated for a few years and then deserted, allowing the natural vegetation to return. The system protects the natural fertility of the soil, and provides it with much time to restore itself after agricultural use. The system works well in areas with low population densities, where most crops are grown for subsistence. With the increase of the population density, the fallow periods are often reduced. This causes a rapid deterioration of the soil and a breakdown of the system. Other disadvantages are the many pests and diseases and the rapid return of the natural vegetation after the first clearing. Some more advanced systems use fertilization and here a gradual transition to a more permanent form of land use is in progress.

A second agricultural system of the tropics is the rice cultivation, as practised in Eastern and Southern Asia. Similar systems are found in parts of Africa and South America, where they have been introduced in some cases by Asians. The system is not limited to the tropics, but is also found in some mid-latitude areas, where summer conditions are similar to those in the tropics. It shows a general adaptation to tropical conditions. In this system, rice is first grown in small seedbeds, and the plants are carefully, one by one, transplanted when they are about 20 cm high. They are moved to flooded fields (paddies, sawahs), which are only dried towards the end of the growing cycle, in the ripening stage. Therefore the supply of water during the critical phases of plant growth is assured, the soil is at the same time well protected against erosion and leaching, weeds are easily kept under control and, when the source of water is a natural stream, nutrients are supplied with it. In other cases fertilization is necessary.

This system is practised by over 400 million people in Asia alone. It allows, because of its intensive use of the land and its high yields, a dense population. The risks involved are low, the main one is that of flooding, which may destroy the small dykes around the fields, which are mostly terraces. Recently the introduction of new varieties of rice has increased yields, in some areas dramatically. In areas where this system prevails, rice is both subsistence and cash crop and the main food.

A third form of agriculture in the tropics is the plantation, a large enterprise of commercial agriculture, introduced in many cases by the colonial powers of their quest for tropical products. Many plantations are now in the hands of the governments of the newly independent states. Plantations typically concentrate on only one crop and include some form of preparation of the products for transportation or storage. Crops such as tea, rubber, oil palm, sisal and sugar cane are largely produced on plantations. Most managers of plantations have a scientific interest in the problems related to their field of production and much research in agroclimatology is carried out in centres maintained by them. The creation of new varieties and the effects of microclimatic and soil conditions are frequently studied here. The results have often helped to reduce losses by diseases and pests and to increase yields. This benefits not only the plantations but also the small farmers who grow the same cash crops.

The long history of mankind in the tropics, and man's ingenuity, have created many different systems of agriculture in the tropics, each of them showing an adaptation to climatic conditions in the area where they are practised (Manshard, 1968, pp. 62–66). Improvements in these systems are not always based on a better, more quantitative evaluation of climate. They often reflect new technical or economic possibilities or changed objectives and attitudes of the tropical farmer. Better communications, for instance, have created improved marketing conditions for many cash crops. New political ideas and progressive national governments have changed the economic attitudes of large segments of the population, of which the main consequence generally is the shift to more cash crops. The introduction of new crops, new strains and entirely new systems demands a new adaptation to climatic conditions, and here the advice of tropical climatologists can be essential.

References

Arkley, R. J., 1963, Relationships between plant growth and transpiration, *Hilgardia*, **34**, 559–584.

Baker, D. N., 1965, Effects of certain environmental factors on net assimilation in cotton, *Crop Science*, **5**, 53–56.

Chang, Jen-Hu, 1968, *Climate and Agriculture*, Chicago, Aldine, 304 pp.

Chang, Jen-Hu, 1970, Potential photosynthesis and crop productivity, *Annals, Association of American Geographers*, **60**, 92–101.

Kenworthy, J. M., 1966, Temperature conditions in the tropical highland climates of East Africa, *East African Geographical Review*, **4**, 4–6.

Manshard, W., 1968, *Einführung in die Agrargeographie der Tropen*, Mannheim, Bibliographisches Institut, 307 pp.

Muturi, S. N., 1968, *Environmental effects on pyrethrins content of pyrethrum*, paper presented at 4th Specialist Committee on Applied Meteorology, Nairobi, November, 1968.

Nieuwolt, S., 1966, A comparison of rainfall during 1963 and average conditions in Malaya, *Erdkunde*, **20**, 180.

Nieuwolt, S., 1973, *Rainfall and evaporation in Tanzania*, B.R.A.L.U.P. Research Paper No. 24, Dar es Salaam, 48 pp.

Nieuwolt, S., 1974, Rainstorm distributions in Tanzania, *Geografiska Annaler*, Series A, **56**, 241–250.

Porter, P. W., 1974, *Potential photosynthesis and agriculture in Tanzania*, B.R.A.L.U.P. Research Paper No. 29, Dar es Salaam, pp. 39–54.

Riehl, H., 1954, *Tropical meteorology*, New York, McGraw-Hill, 93–97.

Strahler, A. N. and A. H. Strahler, 1973, *Environmental geoscience: Interaction between natural systems and man*, S. Barbara, Hamilton, p. 276.

Temple, P. H., and A. Rapp, 1972, Landslides in the Mgeta area, W. Uluguru mountains, Tanzania, *Geografiska Annaler*, Series A, **54**, 165–187.

Thornthwaite, C. W., 1948, An approach toward a rational classification of climate, *Geographical Review*, **38**, pp. 55–94.

Thornthwaite, C. W. and J. R. Mather, 1957, *Instruction and tables for computing potential evapotranspiration and the water balance*, Centerton, N.J., Drexel Institute of Technology, Publications in Climatology, Vol. X, No. 3, p. 244.

United Nations, 1970, *Yearbook of International Statistics*.

United Nations, 1972, *Statistical Yearbook*.

van Rensburg, H. J., 1955, Run-off and soil erosion tests at Mpwapwa, Central Tanganyika, *East African Agricultural Journal*, **20**, pp. 228–231.

Author Index

Academica Sinica, Staff Members, 139, 180
Air Ministry, London, 139, 178
Alaka, M.A., 38, 48
Andre, M.J., 117, 126
Arkley, R.J., 187, 197
Asnani, G.C., 38, 41, 48

Baker, D.N., 187, 197
Balagot, V.F., 143, 178
Barrett, E.C., 169, 178
Barry, R.G., 40, 45, 48, 54, 56, 65, 87, 100
Barton, T.F., 55, 65
Beckinsale, R.P., 35, 48
Bedford, T., 29, 30, 32
Berry, L., 122, 128
Biel, E., 109, 126
Bleeker, W., 117, 126
Blüthgen, J., 29, 32, 71, 82, 83, 113, 126
de Boer, H.J., 120, 126
Bögel, R., 2, 5, 25, 26, 32
Boucher, K., 115, 126
Braak, C., 126, 133, 134, 178
Brier, G.W., 65
Brooks, C.F., 166, 178
Bruzon, E., 139, 178
Byers, H.R., 62, 65, 71, 76, 83, 119, 126

Carton, P., 139, 178
Chang, Jen-Hu, 53, 55, 66, 166, 177, 178, 185, 188, 190, 197
Channa, J.A., 126
Chorley, R.J., 40, 45, 48, 54, 56, 65
Conover, J.H., 98, 101
Conrad, V., 98, 101
Cornford, S.G., 133, 180
Coronas, J., 143, 178
Coyle, J.R., 174, 178
Credner, W., 139, 178
Creutzburg, N., 110, 126, 129, 178
Critchfield, H.J., 25, 30, 32
Crowe, P.R. 46, 48, 116, 122, 126

Dale, W.L., 120, 123, 127
Das, P.K., 55, 56, 65, 135, 138, 178
De Long, G.C. 25, 32
De Ward, R.C., 166, 178

Eliot, J., 135, 178
Espenshade. E.B., 110, 127
Evans, A.C., 116, 127

Finkelstein, J., 117, 127
Fletcher, R.D., 40, 48
Flohn, H., 39, 41, 42, 48, 50, 55, 64, 65, 112, 127, 135, 147, 160, 178
Flores, J.F., 143, 178
Frost, R., 65

García, E., 169, 178
George, C.A., 74, 84
Gilman, Ch. S., 124, 127
Glover, J., 116, 127
Gourou, P., 2, 5
Gregory, S., 68, 76, 83
Griffiths, J.F., 148, 152, 153, 154, 157, 160, 162, 178
Gutnick, M., 165, 178

Hadley, G., 36, 48
Houghton, H.G., 13, 14
Hutchinson. P., 120, 127

Ilesanmi, O.O., 152, 179

Jackson, I.J., 73, 84, 120, 123, 127
Jackson, S.P., 148, 179
James, P.E., 176, 179
Jätzold, R., 129, 160, 179
Jen-Hu Chang, see Chang, Jen-Hu
John, I.G., 53, 66
Johnson, D.H., 41, 48, 126, 127, 158, 160, 179

Kendrew, W.G., 55, 66, 135, 138, 179

Kenworthy, J.M., 26, 32, 116, 127, 186, 197
Khio-Bonthonn, 139, 179
Khromov, S.P., 50, 66
Kimble, G.H.T., 61, 62, 66
Koeppe, C.E., 25, 32
Köppen, W., 1, 4, 5, 17, 32, 110, 127, 129, 179
Koteswaram, P., 42, 48, 54, 55, 66, 74, 84
Kraus, E.B., 117, 127
Krishnamurti, T.B., 42, 48, 70, 84

Landsberg, H., 99, 101
Lautensach. H., 2, 5, 25, 26, 32, 56, 66
Lawes, E.F., 124, 128
Lebedev, A.V., 148, 179
Leroux, M.. 59, 66
List, R.J., 9, 10, 11, 14
Lockwood, J.G., 53, 66, 125, 127, 131, 135, 179
Lumb, F.E., 69, 84, 125, 127

McCallum, D., 125, 127
Malkus, J.S., 41, 48, 77, 84, 117, 127
Manshard, W., 195, 196, 197
de Martonne, E., 110, 127, 129, 179
Mather, J.R., 190, 197
Miller, A.A., 30, 32, 110, 127, 148, 179
Milton, D., 83, 84
Mink, J.F., 79, 84
Mohr, E.C.J., 133, 179
Mörth, H.T., 158, 160, 179
Muturi, S.N., 186, 197

Naval Intelligence Division, London, 139, 179
Newton, C.W., 80, 83, 84
Nieuwolt, S., 16, 23, 26, 31, 62, 76, 84, 95, 101, 127, 133, 141, 159, 179, 189, 190, 192, 197

Orchard, A.Q., 74, 84, 125, 127, 128

Paffen, K., 3, 5, 22, 23
Palmén, E., 36, 48, 80, 83, 84
Pédelaborde, P., 44, 46, 47, 48, 50, 55, 57, 66
Penman, H.L., 88, 101
Petterssen, S., 71, 84, 110, 128
Pollak, L.W., 98, 101
Porter, P.W., 185, 197
Portig, W.H., 169, 179

Ramage, C.S., 50, 51, 62, 66, 68, 69, 75, 76, 84, 117, 119, 128, 131, 134, 140, 179
Rapp, A., 122, 128, 192, 197
Riehl, H., 35, 38, 41, 46, 48, 61, 62, 66, 77, 78, 80, 82, 83, 84, 98, 101, 102, 104, 110, 115, 116, 120, 126, 128, 193, 197
Robinson, P., 116, 127
Rodebush, H.R., 62, 65, 76, 83, 119, 126
Romer, A., 139, 178
Rossby, C.G., 38, 44, 48

Sansom, H.W., 125, 128
Sapper, K., 169, 179
Sasamori, T., 12, 14
Sellers, W.D., 6, 13, 14, 86, 89, 98, 101, 105, 128
Sellick, N.P., 123, 128
Serra, A.B.. 176, 179
Simpson, G.C., 135, 180
Simpson, J., 65
Smith, F.E., 30, 32
Steinhauser. F., 117, 128
Stephenson. P.M., 30, 32
Sternstein, L., 139, 180
Strahler, A.H., 193, 197
Strahler, A.N., 193, 197
Sukanto, M., 134, 180
Sumner, G.B., 74, 84, 125, 126, 127, 128

Taylor, C.M., 123, 124, 128
Temple, P.H., 122, 128, 192, 197
Terjung, W.H., 30, 32
Thom, E.C., 30, 32
Thompson, B.W., 118, 119, 128, 148, 150, 178, 180
Thornthwaite, C.W., 85, 88, 101, 110, 128, 129, 147, 180, 188, 190, 197
Tison, L.J., 87, 101
Torrance, J.D., 120, 128
Trewartha, G.T., 25, 26, 27, 28, 32, 152, 180
Troll, C., 27, 33, 101, 128, 129, 180

United Nations, 182, 197

Van Den Berg, C.A., 166, 180
Van Loon, H., 43, 48
Van Rensburg, H.J., 192, 197
Vulquin, A., 176, 180

Watts, I.E.M., 62, 66, 69, 74, 76, 84, 94, 100, 101, 119, 125, 126, 128, 134, 138, 180
Webb, C.G., 30, 33
Weischet, W., 107, 128, 172, 180
Wexler, R., 62, 66, 103, 128
Wiesner, C.J., 87, 101, 104, 128
von Wissmann, H., 110, 128, 129, 180

Woo Kam Seng, S., 120, 128
Wüst, G., 87, 101
Wycherley, P.R., 120, 128

Yin, M.T., 54, 66,

Zobel, R.F., 133, 180

Geographical Index

Accra, Ghana, 122
Addis Ababa, Ethiopia, 114, 115, 159, 160
Africa, 57–59, 129, 147–162, 196
    East Africa, 59, 89, 112, 114, 119, 148, 149, 157–160, 195
    equatorial Africa, 148, 150–152
    North Africa, 89, 100
    southern Africa, 59, 89, 147, 149, 154–157
    West Africa, 57–59, 75, 76, 78, 89, 100, 105, 109, 150–154
Agadès, Niger, 153, 154
Agulhas Current, 17
Amazonas Basin, 107, 173–175
Amboina, Indonesia, 132
Andagoya, Colombia, 170, 171
Andes Mountains, 89, 162–164, 169, 171–173
Annam, 140
Aparri, Philippines, 141–143
Arabian Sea, 69, 75, 82, 83
Asia
    eastern Asia, 100, 129, 130, 196
    southern Asia, 89, 100, 129, 138, 147, 168, 173–175, 196
Atlantic Ocean, 36, 57, 82, 83, 100, 105, 129, 151, 153–155, 163
Australia, 50, 52, 53, 56, 57, 79, 83, 89, 129, 143, 144, 146, 147

Bahamas, 166
Bandung, Indonesia, 26, 132
Bangladesh. 76, 134–138, 182
Battambang, Cambodia, 191–192
Bay of Bengal, 54, 74, 82, 83, 135, 136, 138
Beira, Mozambique, 16, 156, 157
Beitbridge, Zimbabwe, 91
Belem, Brazil, 114, 115
Belize, 167, 168

Benguela Current, 21, 147, 151, 152, 155, 157
Bijapur, India, 88
Bogor, Indonesia, 133
Bolivia, 182
Bombay, India, 19, 122, 135, 136, 138
Borneo, Indonesia, 16, 105
Botswana, 156, 157
Brazil, 170, 173–176, 195
Brazzaville, Congo, 156, 157
Bujumbura, Burundi, 156
Bulawayo, Zimbabwe, 91
Burma, 19, 53, 55, 74, 105, 122, 134, 138–141

Calcutta, India, 91, 115, 122, 135, 136, 138
California Current, 169
Callao, Peru, 170, 172
Cambodia, 191, 192
Cameroons. 93, 105, 115
Canaries Current, 147, 154
Canton, China, 88
Canton Island, Pacific Ocean, 176, 177
Caracas, Venezuela, 114, 115
Carnarvon, W. Australia, 16, 17
Caribbean, 34, 77–79, 80, 83, 105, 129, 164–168
Central America, 107, 129, 164, 166–169
Chile, 95
China, 52, 53, 54, 56, 88, 112
Colombia, 164, 170, 171, 182
Colombo, Sri Lanka, 191, 192
Colón, Panama, 167, 168
Congo Basin, 147, 150–152, 156, 158
Congo Republic, 156, 157
Costa Rica, 167–169
Cuba, 166
Cuiabá, Brazil, 173, 175
Curaçao, West Indies, 175

Dakar, Senegal, 153, 154
Dar es Salaam, Tanzania, 91, 118, 124
Darwin, Australia, 143
Deccan Plateau, India, 137, 138
Djakarta, Indonesia, 26, 122, 191, 192
Djibouti, Afars & Issas, 93
Dodoma, Tanzania, 191, 192
Durban, South Africa, 16, 17

Eala, Zaire, 151, 152
East African Rift Valley, 57, 125, 157, 158
Ecuador, 122, 172, 173
El Salvador, 122
Entebbe, Uganda, 91, 122, 151, 152
Enugu, Nigeria, 88, 114, 115
Erigavo, Somalia, 159, 160
Ethiopia, 21, 114, 115, 153, 159, 160

Florida, USA, 76, 115, 119, 164–166
Fort de France, Martinique, 165, 166
Fraser's Hills, Malaysia, 26

Ganges Lowlands, 53, 64, 105, 135, 137, 145
Georgetown, Guyana, 122
Ghana, 122
Ghats, India, 137
Guam, Pacific Ocean, 78
Gulf of Guayaquil, Ecuador, 171
Gulf of Tonkin, 140
Guyana, 122, 174

Hawaii, 69, 79, 176, 177
Hilo, Hawaii, 176, 177
Honduras, 168, 182
Hong Kong, 122, 191, 192
Honolulu, Hawaii, 176, 177
Humboldt Current, 21, 163, 169, 171
Hyderabad, India, 191, 192

Iloilo, Philippines, 141, 142
India, 19, 26, 53–55, 69, 74, 76, 88, 91, 105, 108, 112, 115, 122, 134–138, 144–146, 182, 191, 192, 195
Indian Ocean, 55, 56, 59, 83, 107, 129, 146, 155, 156, 161, 176–178
Indo-China, 53, 54, 134, 138–141
Indonesia, 26, 52, 53, 56, 100, 105, 122, 131–134, 143–144, 182, 191, 192
Iran, 21, 144
Ivory Coast, 152, 182

Jamaica, 165, 166
Japan, 50, 52, 53, 55, 56, 79
Java, Indonesia, 131
Jodhpur, India, 145, 146

Kabwe, Zambia, 19
Kananga, Zaire, 151, 152
Karachi, Pakistan, 145, 146
Kasama, Zambia, 88
Kenya, 26, 27, 91, 108, 118, 124, 125, 158–160
Khartoum, Sudan, 19
Kingston, Jamaica, 165, 166
Korea, 52, 53, 56
Koshun, Taiwan, 91
Kota, India, 135, 136, 138
Kota Bharu, Malaysia, 132
Kribi, Cameroons, 93
Kuala Lumpur, Malaysia, 26, 88
Kupang, Timor, 16, 143

Lagos, Nigeria, 122, 153, 154
Lake Kariba, 96
Lake Nyasa, 158
Lake Victoria, 62, 91, 119, 150, 152, 157, 158
Laos, 114, 115, 182
Legaspi, Philippines, 141, 142, 143
Liberia, 105, 151, 152
Lima, Peru, 170–172
Lindi, Tanzania, 159–160
Livingstone. Zambia, 108
Lourenço Marques (= Maputo), Mozambique, 115
Luang Prabang, Laos, 139–140
Lusaka, Zambia, 24, 25, 31, 32, 92, 93, 114, 115, 191, 192

Madagascar, 59, 91, 105, 114, 147, 149, 150, 156, 157, 160–162
Madras, India, 135, 137
Maintirano, Malagasy, 91
Majunga, Malagasy, 161, 162
Malacca, Malaysia, 76
Malawi, 182
Malaya, 26, 52, 53, 56, 76, 88, 89, 91, 96, 114, 115, 118, 119, 131–134, 138
Mali, 19
Mananjary, Malagasy, 161–162
Mangalore, India, 135, 137, 138
Manila, Philippines, 141, 142
Maputo (= Lourenço Marques), Mozambique, 115

Maracaibo, Venezuela 170, 175
Martinique, West Indies, 165, 166
Masulipatnam, India, 136–138
Maun, Botswana, 156, 157
Mauritius, Indian Ocean, 114, 115
Mazatlán, Mexico, 168, 169
Merauke, New Guinea, Indonesia, 143, 144
Mergui, Burma, 139, 140
Mérida, Mexico, 168, 169
Mexico, 21, 27, 167–169
Mexico City, 27, 168, 169
Miami, Florida, 115, 165, 166
Mogadiscio. Somalia, 158–160
Mombasa, Kenya, 27
Monrovia, Liberia, 151, 152
Morocco, 95, 100
Moyale, Kenya, 91
Mount Cameroon, 105
Mozambique, 16, 115, 156, 157
Mozambique Channel, 155, 162

Nagpur, India, 108
Nairobi, Kenya, 26, 108, 118, 124, 125, 158–160
Nakhon Ratchasina, Thailand, 139, 140
Namibia, 95, 100
New Delhi, India, 26
New Guinea, 107, 131–134, 143, 144
Niamey, Niger, 93
Niger, 93, 153, 154
Nigeria, 88, 114, 115, 122, 152–154

Oahu, Hawaiian Islands, 79

Pacific Ocean, 36, 39, 52, 53, 69, 77, 80, 82, 83, 100, 105, 107, 129, 164, 168–171, 176–178
Pakistan, 50, 144, 182, 195
Panama, 164, 166–168, 171
Patna, India, 19
Penang, Malaysia, 114, 115
Peru, 95, 100, 170–172, 182
Peru Current, 163, 169, 171
Philippines, 56, 79, 141–143, 182
Pontianak, Borneo, 16
Port Hedland, Australia, 145, 147
Port Swettenham, Malaysia, 76
Puerto Rico, 122, 165, 166

Queensland. Australia, 144
Quito, Ecuador, 122, 172, 173

Rangoon, Burma, 19, 122
Remanso, Brazil, 173, 175

Sahel, West Africa, 109, 154
Sandakan, Malaysia, 132
Salina Cruz, Mexico, 167–169
San José, Costa Rica, 167, 169
San Juan, Puerto Rico, 122, 165–166
San Salvador, El Salvador, 122
Santa Elena, Brazil, 170, 173, 176
Santarém, Brazil, 173, 175, 176
Senegal, 153, 154
Seychelles, 16, 17
Sierra Leone, 105
Simla, India, 26
Singapore, 24, 25, 31, 32, 64, 76, 89, 91–93, 122–124, 190–191
Somalia, 147, 158–160
South America, 60, 89, 105, 114, 129, 162, 169–176, 196
South China Sea, 34, 52, 53, 139, 140, 141
Sri Lanka, 74, 134–138, 191, 192
Straits of Malacca, 62, 76, 119
Sudan, 19, 68, 148, 152–154
Sumatra, 76, 105

Tabora, Tanzania, 91, 159, 160
Taiwan, 91
Tamatave, Malagasy, 91
Tampico, Mexico, 168, 169
Tanah Rata, Malaysia, 89, 91
Tananarive, Malagasy, 161, 162
Tanzania, 91, 118, 158–160, 182, 183, 191, 192
Tennant Creek, Australia, 145, 147
Thailand, 134, 138–141, 182
Thar Desert, 68
Tibetan–Himalayan massif, 53, 54, 55, 64
Timbuktu, Mali, 19
Timor, 16
Tourane, Vietnam, 139, 140
Townsville, Australia, 143
Trincomalee, Sri Lanka, 135, 137
Trivandrum, India, 135, 137, 138
Truk, Caroline Islands, 176, 177
Tuléar, Malagasy, 161, 162

Uganda, 91. 122, 151, 152

Venezuela, 114, 115, 166, 170–175

204

Vientiane, Laos, 114, 115
Vietnam, 139, 140

Wau, Sudan, 153, 154

Yaounde, Cameroons, 114, 115

Yucatan, Mexico, 168

Zaire, 151, 152
Zambia, 19, 24, 25, 31, 88, 96, 108, 114, 115, 191, 192
Zimbabwe, 91

Subject Index

agriculture, tropical, 4, 148, 182–197
agroclimatology, 183, 196
albedo, 8, 9
altitudinal zones, 27, 28
anabatic wind, 63, 64
anti-trades, 36, 42, 43

Bergeron effect, 103
breezes, 60–63, 100, 162
burst, monsoon, 136

cloudiness, 11, 23, 98–100, 184
clouds, 94–98
coalescence, 103
condensation, 94, 95
Congo Air Boundary, 155, 156
continentality, 23, 149
convection, 68, 76, 87, 103, 131, 133
convergence, 69, 70, 95, 98, 103, 131, 144, 152
Coriolis force, 41, 44, 63, 67, 69, 82
crachin, 140
cyclone, tropical, 79–83, 138, 140, 142, 165, 169

dew, 92, 94
Discomfort Index, 30
disturbance line, 75, 154
disturbances, 67–83, 95
divergence, 157, 159–160
doldrums, 40, 41

easterly wave, 47, 77–79, 95
elevation, 25–28, 160, 169, 172–173, 186
    of the sun, 8, 9
El Nino, 171
eluviation, 193
energy flow, 14
equator, meteorological, 39
equatorial bridge, 158
equatorial trough, *see* Inter Tropical Convergence Zone

Et/E_0 ratio, 188, 189
evaporation, evapotranspiration, 85–88, 188–192
exposure, 160–162, 169, 172–173, 176, 194
eye, of tropical cyclone, 80

fog, 94, 95

Ganges-type, of temperature regime, 17, 19, 136, 144, 146
garúa, 172

Hadley cell, 35–38
harmattan, 57
heat
    latent, 13, 67, 85
    sensible, 13
heat lows, 68
high, subtropical, 36, 37, 42–45, 166, 170, 171
highlands, 1, 3, 21, 26–28, 89, 120, 149, 185
humidity, 29, 88–94, 187
hurricane, 79–83, 165, 169

infiltration, 192
insolation, 6–11, 29, 87, 88, 183–186
instability, 67, 68, 71, 76
Inter Tropical Convergence Zone, 36, 38–42, 47, 75, 87, 98, 100, 111
    in Africa, 59, 152–156, 158–159
    in Madagascar, 160–162
    in South America, 163, 169–171, 173, 175
    in the Caribbean area, 165, 168–169
    over oceans, 105
Inter Tropical Discontinuity, 59
inversion, trade wind, 45–47, 78

Jet Stream, 42, 55

katabatic wind, 63–65

lake breeze, 62
land breeze, 61–63, 76, 100, 119
leaching, 193
linear system, 69, 75–76, 98
long rains, 158

mango rains, 140
maritime continent, 131
monsoon depression, 74, 75
monsoons, 49–60
  in Africa, 57–59, 153–154, 157–160
  in Asia, 50–56, 131, 136–146
  in Australia, 56–57, 143–144
  in South America, 60, 162, 174
morning fog, 94, 95
mountain wind, 63–64

oasis effect, 87
ocean currents, 13, 17, 45, 147, 162–163, 171
ocean water surface temperature, 80, 89, 147, 171, 172
orographic lifting, 76, 103, 105, 107, 131, 157

paramó belt, 28
pests, 187, 195
photoperiodism, 186
photosynthesis, 183–186, 195
polar front, 53, 69, 95, 139–140, 164
precipitation, see rainfall
probable maximum precipitation, 125

radiation
  solar, see insolation
  terrestrial, 11–14
radiation balance, 13, 14
rainfall, 102–126
  diurnal variation, 116–119
  frequency, 119–120
  in monsoon depression, 74, 75
  in thunderstorm, 73–74
  in tropical cyclone, 83
  intensity. 125–126, 133, 192
  orographic, 64, 107
  seasonal variation, 109–116
  variability, 56, 107–109, 160
rainforest, 87
rainstorm, 125–126
reflection, 11, 85
respiration, 184, 185

saturation point, 94
sea breeze, 61–62, 97, 100, 119
  front, 119
shifting cultivation, 195, 196
short rains, 158
snowline, 172
soil erosion. 122, 195
solar constant, 6
solar energy, 181
solar radiation, see insolation
spells, dry and wet, 120, 121, 152–154
spring rains, 160
squall-line, 71, 75–78
stability, 85
stratiform cloud, 95, 98
sultriness, 29
sumatras, 76, 131

temperature, 15–32
  effective, 30
  physiological, 15, 28–32
  sea-level. 2,3
Temperature–Humidity–Index, 30
thermal equator, 20, 21
thermal lows, 49
thunderstorm, 68, 70–74, 95, 103, 133
tierra caliente, 27
tierra fría, 28
tierra helada, 28
tierra templada, 27
trade winds, 36, 37, 45–47, 53, 161–162, 171, 172
  confluence, see Inter Tropical Convergence Zone
  disturbances, 75, 77–79
  in North Atlantic, 164–168, 171, 173
  in North Pacific, 141–142
  in South Atlantic, 173–175
  in South Pacific, 144–146
  inversion, 45–47, 78, 164–166
transpiration, 85, 187, 188
tropics, 1–4
  dry, 3, 4, 86, 130, 144–147, 177, 185, 194
  humid, 2, 3, 185
tropopause, 12, 36
troposphere, 12, 34
turbulence, 86
typhoons, 79–83, 140, 142, 144

uniformity, 15–20, 134

valley wind. 63–65

ventilation, 29
vernalization, 186
vorticity, 80, 97

warm rain, 103
water balance, 188–192
water vapour, 85–94

wave, easterly, 47, 77–79, 95
willy-willies. 79–83, 144
wind power, 181
wind shear, 82, 98
winds, local, diurnal, 60–65

zenithal rains, 111